Freshwater Dinoflagellates of North America

Freshwater Dinoflagellates of North America

SUSAN CARTY

COMSTOCK PUBLISHING ASSOCIATES
a division of
Cornell University Press
ITHACA AND LONDON

First published 2014 by Cornell University Press
Printed in the United States of America

Library of Congress Cataloging-in-Publication Data

Carty, Susan, 1948– author.
  Freshwater dinoflagellates of North America / Susan Carty.
    pages cm
  Includes bibliographical references and index.
  ISBN 978-0-8014-5176-8 (cloth : alk. paper)
  1. Dinoflagellates—North America—Classification. 2. Freshwater
algae—North America—Classification. I. Title.
  QK569.D56C37 2014
  579.8′7—dc23   2013035403

Cornell University Press strives to use environmentally responsible suppliers and materials to the fullest extent possible in the publishing of its books. Such materials include vegetable-based, low-VOC inks and acid-free papers that are recycled, totally chlorine-free, or partly composed of nonwood fibers. For further information, visit our website at www.cornellpress.cornell.edu.

Cloth printing   10 9 8 7 6 5 4 3 2 1

Contents

Thecate/Armored Taxa

Color plates follow page 116

Preface

Dinoflagellates are the last group of algae I learned. Compared with other phyla, they seemed to have the least morphology (the round brown cells)— but as I discovered, they are enormously interesting. They have a rich fossil history, there are fantastically shaped marine species, some dinoflagellates are toxic, some species are the zooxanthellae of reef building corals, and some are bioluminescent. They are unified by a suite of characters that sets them apart from all other algal groups and makes finding their nearest cousins difficult. They have varied shapes and varied nutrition (photosynthetic, mixotrophic, and phagotrophic), they are motile or immobile, thecate or athecate. I have been studying dinoflagellates since 1981, and even though they require a microscope for identification, I consider myself a field biologist. I sample freshwater everywhere I go and have examined thousands of samples.

What this book is: it is my goal to make dinoflagellates less intimidating. I hope to make their identification easier for people looking at water samples. I hope, by showing how much we do not know about freshwater dinoflagellates, that people will begin to ask questions and try to find answers. I am doing this by writing descriptions from my own experiences and from the descriptions of others who have seen the species. I begin each description with a diagnosis, which includes salient features. Descriptions are supplemented by line drawings that call attention to features to look for. Where possible, I include light and scanning electron micrographs of the species to help make the connection between what I have seen in my microscope and what you are seeing in yours. This is a compendium of species reported from North America, but I have updated nomenclature to include likely taxa and questionable species. Distribution maps are included to highlight how little we know about most species. In addition to figures located throughout, this book includes black-and-white plates (indicated by PL) and color plates (CP).

What this book is not: a treatise on the biology of freshwater dinoflagellates. Several excellent books have been written for the reader intrigued by this interesting group of protists, particularly those by Spector (1984), Taylor (1987), Evitt (1985), and Fensome et al. (1993). I have tried to focus on making freshwater dinoflagellates more accessible; they are poorly reported from lakes (I know they are there!) because of the difficulty in identifying them. I am aware that the nomenclature will be outdated before this is printed, but it is correct as of 2012. I hope this book will inspire others to look for dinos, work out life histories, get them into culture, and use them in molecular analyses.

Acknowledgments

A work like this cannot be completed in a vacuum; many people have contributed and I greatly appreciate their help.

Victor W. Fazio, my longtime collecting buddy, and I have traveled thousands of miles and visited hundreds of ponds, lakes, and reservoirs, he for birds and dragonflies, I for algae. Nancy Rubenstein, reference librarian and now director of Beegley Library at Heidelberg University, did a tremendous job of tracking down original species descriptions; this work could not have been done without her. John Hall, a former student and a lifelong colleague, traveled and collected specimens with me, shared adventures, continues to send images and references, and generously agreed to be a reader and editor of the manuscript. My friend Mary Ann Rood provided editing, moral support, and several illustrations. Elenor Cox, Greta Fryxell, and Lois Pfiester, my mentors, have supported me and have always provided encouragement. Several other readers and editors and two anonymous reviewers contributed to a better book.

I am grateful for the support of many people at Heidelberg University: my colleagues in the Department of Biological and Environmental Studies, Ken Baker, Amy Berger, Pam Faber, Ken Krieger, director of the National Center for Water Quality Research, Kylee Spencer, and Katie Wise; and students Tara Ensing, Maggie Hess, Ben Laubender, and Brian Sutherland. Heidelberg-awarded Aigler grants and sabbaticals were instrumental in my research for and writing of this book.

Freshwater
Dinoflagellates
of North America

Introduction

When I cast my plankton net into a lake from the shoreline, nearby fishermen look at me wonderingly until I explain that, while they are sampling the top of the food chain, I am sampling the bottom. Dinoflagellates are a type of algae, and algae are the single-celled aquatic organisms that harvest the sun's energy in photosynthesis and provide food to everything bigger than they are. In addition to providing food, dinoflagellates produce blooms that can give a foul smell and taste to water or even cause fish kills. It is these understudied freshwater dinoflagellates that are the subject of this book. My hope is that it will stimulate interest in freshwater dinoflagellates and provide a resource not just for researchers and ecologists who hope to document algal populations in their lakes and streams, but also for the people who manage the water we drink and the lakes we use for recreation.

What those puzzled fishermen most likely have heard of is toxic red tides that contaminate fish and shellfish. Dinoflagellates have a well-known and well-deserved reputation for toxicity. Estuarine dinoflagellates are responsible for red tides and considered one of the primary groups responsible for harmful algal blooms (HABs) that cause toxic shellfish poisoning (TSP), paralytic shellfish poisoning (PSP), ciguatera, and fish kills. They can produce potent neurotoxins and hemolytic compounds. Toxin-producing marine and estuarine dinoflagellates have been extensively studied in an effort to understand the environmental conditions that trigger blooms and their consequences. The fish and shellfish industries along the Atlantic coast of North America are particularly sensitive to dinoflagellate blooms, and these blooms may intensify with global warming (Hallegraeff 2010).

Although freshwater dinoflagellates have only a few species documented to produce toxic effects (see "Toxins" under "Dinoflagellate Biology"), for the most part their toxicity has not been studied (see future research needed). The potential for danger certainly exists. Freshwater dinoflagellates are capable of producing huge blooms, and some do so in water supply reservoirs, sometimes causing taste and odor problems (Knappe et al. 2004). The genetic similarity of marine and estuarine species to freshwater species is rarely studied (Logares et al. 2007a,b), but it might be important for understanding morphological trends and potential for toxin production.

Perhaps unexpectedly, fossil dinoflagellates are more thoroughly studied than their living counterparts. When I began my studies on dinoflagellates, I was encouraged by colleagues working with fossil dinoflagellates. At that time, they were all employed by oil companies, since dinoflagellate cysts (dinocysts) are important in stratigraphy. Recent meetings of the international dinoflagellate group, however, have focused on fossil dinoflagellates as indicators of world climatic conditions, including past global warming events. This climate work relies entirely on fossil marine dinocysts, but freshwater species also produce resistant cysts, and their presence in sediments can indicate coastal deposition.

FUTURE RESEARCH NEEDED

It is my goal for this book to stimulate interest in freshwater dinoflagellates. Much basic information about these algae is still unknown, including the following:

Life Cycles

Relatively few complete life cycles for freshwater species have been worked out. Dinoflagellates, especially the freshwater parasitic species, can have quite complex life cycles. It is possible that some dinoflagellate species may be only stages in another's life cycle (hinted at for *Dinastridium* and *Hemidinium/Gloeodinium*). In addition, the cyst stage of many dinoflagellates is unknown. Understanding and identifying dinoflagellate cysts is important because they may serve as the seed for future communities, including blooms, and because the cysts are the most readily fossilized stage of the life cycle. The ability to identify cysts would provide a valuable link between living and fossil algal communities.

Biogeography and Ecology

Researchers using fossil marine dinocysts to plot global warming events and continental movement can do so because certain species are indicators of particular conditions. So little is known about ecological requirements of individual freshwater species that only a few species can be labeled as cold loving, found in acidic environments, etc. Our knowledge of the distribution and preferred environmental conditions of North American species is limited. Some species are common and can be found almost everywhere (*Ceratium*); others might be rare or merely unreported. As distribution maps accompanying species descriptions were compiled for this book, it became clear that some provinces, states, and countries are well studied and others are not, so there is much to be learned about species distributions and the conditions under which they are found.

Unusual Habitats

Following this introductory section are keys to genera. The first is a classical morphological key; the second a key based on habitat that immediately separates out taxa not found free swimming or floating in the plankton. Non-planktonic dinoflagellates are an area rich for exploration. For example, only two species attached to fish are reported from North America, though fish parasites are well known in freshwater and marine habitats. Almost 150 species of dinoflagellates infect invertebrates (Levy et al. 2007), though none have been reported from North America. Protists are also known to have dinoflagellate parasites, but this work has been confined to marine ciliates (Coats et al. 2010). Recent work on algae associated with aquatic carnivorous plants recorded one dinophyte (Wołowski et al. 2011). In addition to investigations into the possibility of parasitic/epizooic freshwater dinoflagellates, exploration is needed into the entire area of symbioses. Dinoflagellates are the famous endosymbionts of reef building corals, but there is no documentation of endosymbiosis with freshwater protozoa or invertebrates (Reisser 1992, Reisser and Widowski 1992). Benthic habitats should also be routinely collected. So far there is only one report of one dinoflagellate found in freshwater sand. Many species of marine sand-dwelling dinoflagellates are documented. Terrestrial habitats are another possibility. Cliff faces with continuous wetting (drip walls) may harbor the quasi colonial *Rufusiella*. Many other terrestrial habitats have associated algae, for example soil crusts (cyanobacteria and diatoms) and tree trunks (green algae), that should be examined for dinoflagellates. It is likely that species new to science will be discovered with a more thorough investigation of unusual habitats.

Toxicity

Although much research focuses on toxicity of marine and estuarine species (see proceedings of the annual International Conference on Harmful Algae), few freshwater dinoflagellate species have been tested for selective toxicity. Considering the large number of species capable of forming blooms (see Table 2), more work needs to be done.

Phylogeny

Few freshwater dinoflagellates have been analyzed using morphological, anatomical, and molecular methods to determine phylogenetic relationships to marine taxa or to each other (Logares et al. 2007a). Some species of morphologically identical freshwater dinoflagellates have international distributions (*Ceratium hirundinella, Peridinium gatunense*). Freshwater habitats are islands, separated by land and salt water, where evolution continues, and molecular analyses of species with worldwide distributions might provide insights into genetic variability within and between morphological species.

WHAT IS A DINOFLAGELLATE?

Dinoflagellates are microscopic algae that are different from other algae. As I look through the microscope I can see that the distinctive appearance of dinoflagellates is the result of several features: golden color; an indented waist called the *cingulum* that houses a flagellum encircling the cell; and spiral swimming pattern due in part to the encircling flagellum. Some collectively unique features of cells include storing starch (which turns black if you add iodine), a large nucleus with permanently condensed chromosomes, and ejectile organelles called *trichocysts*.

HISTORY OF NORTH AMERICAN FRESHWATER DINOFLAGELLATE RESEARCH

"Several years ago the writer, attempting to identify the dinoflagellates occurring in fresh-water plankton collections from various parts of the United States, found that the literature on this group was difficult of access, old, and widely scattered. Practically nothing has been written on this group in the United States," Samuel Eddy wrote in 1930, in a paper titled "The fresh-water armored or thecate dinoflagellates." Not much has changed since then. His was the first work to focus on North American occurrences of freshwater dinoflagellates; he reviewed 7 genera and 55 species, including comments on American occurrence of 5 genera (*Hemidinium, Glenodinium, Gonyaulax, Peridinium, Ceratium*) and 16 species. In addition to referring to European workers on dinoflagellates (Schilling 1891a, 1913, Lemmermann 1910, Wołoszyńska 1913, 1916, 1917, Lindemann 1925), Eddy consulted the works of Playfair (1912, 1919), working in Australia. He referred to North American workers Kofoid (1907, 1909), on marine taxa, and Prescott (1927, 8 dinoflagellate species in Iowa). Of the 16 species Eddy reported as occurring in the United States, he had seen 15, including one new to science, *Peridinium wisconsinense*.

Wailes (1934) referred to Eddy's work, extended it to include Alaska and Canada, included records not found in Eddy, and reported 49 species and varieties for North America. He included athecate genera (*Amphidinium* and *Gymnodinium*) and added *Peridiniopsis* and *Kryptoperidinium* (a genus no longer used for freshwater species) to the list of thecate genera known to occur in North America. Included were three new species: *Peridinium cucumis, P. kincaidi,* and *P. vancouverense* (Wailes 1934).

Although no one has assembled a species list since 1934, there have been researchers who studied freshwater dinoflagellates, phycologists who included dinoflagellates (with descriptions and illustrations) in regional species lists, and freshwater biologists and ecologists who included them in phytoplankton lists.

Rufus Thompson wrote three important papers dealing exclusively with freshwater dinoflagellates. The first, "Fresh-water dinoflagellates of Maryland" (1947), began by describing methods (squeezing macrophytes, using light to attract motile cells) that have been useful to later investigators. He mentioned 21 species of freshwater dinoflagellates previously reported from the United States; to this he added 10 species, 2 new to science. One of the species he reported, *Glenodinium quadridens,* was renamed *Peridiniopsis thompsonii* by Bourrelly (1968a), since the cell as illustrated did not conform to *P. quadridens.* The second paper, "Immobile Dinophyceae" (1949), focused on a group not previously reported, discussed 7 species, one new to science, with detailed observations on reproduction. The third paper, "A new genus and new records of freshwater Pyrrhophyta in the Desmokontae and Dinophyceae" (1950), based principally on work done in Kansas, established the genus *Woloszynskia* and a new species, *W. reticulata,* and described new species in other genera. A species described in the paper, *Peridinium intermedium,* was later moved to a new genus *Thompsodinium,* named by Bourrelly (1970) in honor of Thompson. Another species in the paper, *Glenodinium ambiguum,* was the basis of the new genus *Kansodinium* (Carty and Cox 1986). One of Thompson's graduate students (Richards 1962) worked on the life history of *Urococcus insignis* (originally in the Chlorophyta), later renamed *Rufusiella* by Loeblich III (1967) to honor Thompson. Of the 20 taxa Thompson reported from Kansas, 2 new species became the basis for new genera, a new genus was described, and 3 other species new to science were described. I have relied heavily on Thompson's descriptions and illustrations for taxa I have not seen.

Lois Pfiester began her career with *Ceratium hirundinella* (1971), then worked elucidating the life histories of dinoflagellates. She had a gift for isolating cells into culture and the patience to follow and document their sexual life histories when sexuality in dinoflagellates seemed sporadic. While at the University of Oklahoma, she published on the sexual reproduction of *Peridinium cinctum* f. *ovoplanum* (1975), *P. willei* (1976), and *P. gatunense* (1977). With coauthors, she investigated *Hemidinium nasutum* (Pfiester and Highfill 1993), *Cystodinium bataviense* (Pfiester and Lynch 1980), *Woloszynskia reticulata* (Pfiester et al. 1980), *Peridinium volzii* (Pfiester and Skvarla 1979), *P. limbatum* (Pfiester and Skvarla 1980), and *P. inconspicuum* (Pfiester et al. 1984). The cultures established by Pfiester enabled other types of investigation into dinoflagellates: counting chromosomes (Holt and Pfiester 1982), auxotrophy (Holt and Pfiester 1981), electrophoresis of soluble enzymes (Hayhome and Pfiester 1983), hypnozygote cell layers (Timpano and Pfiester 1986), and ecology and excystment (Chapman and Pfiester 1995). Pfiester collaborated with Jiri Popovský on several papers and on the widely used volume Dinophyceae in *Süßwasserflora von Mitteleuropa* (Popovský and Pfiester 1990). She is immortalized in the genus *Pfiesteria* (Steidinger et al. 1996a). Jiri Popovský also published on thecate dinoflagellates in Cuba (1970) and coauthored several papers with Pfiester in which her material was from Oklahoma and his was from Czechoslovakia (Pfiester and Popovský 1979, Popovský and Pfiester 1986).

General algal references of specific regions that include dinoflagellates and contain illustrations by the author(s) include "The motile algae of Iowa" (Prescott 1927), *Algae of the Western Great Lakes* (Prescott 1951b), *The algae of Illinois* (Tiffany and Britton 1952), *Handbook of algae with special reference to Tennessee and the southeastern United States* (Forest 1954a), *A manual of fresh-water algae in North Carolina* (Whitford and Schumacher 1969), and *The algae of western Lake Erie* (Ohio; Taft and Taft 1971). Along with the works of Eddy (1930), Wailes (1928a,b, 1934), and Thompson (1947, 1949, 1950), these make up the major sources of author-illustrated dinoflagellate reports, and some of their illustrations are included in the present work. Illustrations are critical for species identification. If an author has accurately illustrated the species seen, the correct name can be attached (see *Peridiniopsis cunningtonii, P. quadridens, P. thompsonii*). Some of the species, and most of the location references for species, included in this work are from species lists without illustration. I am grateful to the authors who have compiled regional lists of species, including those of the southeastern United States (Dillard 2007), and checklists of Canadian provinces including the maritimes (Hughes 1947–1948), British Columbia (Stein and Borden 1979), Labrador (Duthie et al. 1976), Northwest Territories (Sheath and Steinman 1982), and Ontario (Duthie and Socha 1976).

My own regional contributions began with thecate dinoflagellates from Texas (1986), and have extended to Ohio (Carty 1993), Ecuador (Carty and Hall 2002), Belize (Carty and Wujek 2003), Alaska (Carty 2007), and New England, New Brunswick, and Nova Scotia (Carty 2009). In other papers I have reviewed the taxonomy of various taxa: *Lophodinium* (Carty and Cox 1985), *Kansodinium* and *Durinskia* (Carty and Cox 1986), *Thompsodinium* (Carty 1989), and *Parvodinium* (Carty 2008). I also wrote the chapter on dinoflagellates for *Freshwater algae of North America* (Carty 2003), which provides a key to genera and was an impetus to this work.

In the 80 years since Eddy and Wailes, there have been no further compilations of species until this work. Eddy focused on the United States, and Wailes on Canada (primarily British Columbia); it is my intention to include species reported from Mexico, Central America, the Caribbean, and Greenland.

DINOFLAGELLATE BIOLOGY

The Cell

Dinoflagellates are eucaryotic unicellular protists that can be found in freshwater, estuarine, and marine habitats. They can be recognized by the presence of a *cingulum* (a pinched-in waist, the transverse groove), or golden color, or both. The cingulum houses a tinsel *flagellum* whose beating causes the cell to rotate (*dino* in *dino*flagellate and *-dinium* in many genus names refer to this whirling). Cells also have a longitudinal groove called the *sulcus*, which directs a whiplash flagellum posteriorly that propels the cell forward. These two dissimilar flagella and their placement on the cell are unique to dinoflagellates (Fig. 1). The flagella also impart a distinctive swimming pattern, whirling while moving forward, that forms a spiral (Fenchel 2001). This unusual swimming pattern can be used to recognize dinoflagellates in a mixed sample.

Dinoflagellates exhibit a large range of nutrition modes. The term *assimilative*, rather than *vegetative* (implying plantlike), will be used here for the stage that takes in nutrients. Photosynthetic cells may have their own chloroplasts, harbor photosynthetic endosymbionts, or use plastids stolen from prey (kleptoplastids). *Heterotrophic* dinoflagellates may be predaceous, parasitic, or osmotrophic; *mixotrophic* species use both photosynthesis and heterotrophy.

The golden color in most photosynthetic dinoflagellates is due to a unique carotinoid pigment, peridinin, in the chloroplasts. The color can range from yellow to deep brown (see color plates). Dinoflagellate chloroplasts are typically triple membrane bound with thylakoids in threes and starch stored outside the chloroplast (Dodge 1975). Typical dinoflagellate chloroplasts with peridinin and chlorophyll c2 are widespread in the phylum and considered to have been acquired early in the evolution of the group (Saunders et al. 1997).

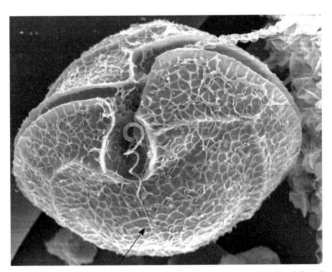

Figure 1. Two dissimilar flagella on *Peridinium gatunense:* the longitudinal flagellum (arrow) and transverse flagellum (loosened from cingulum).

Another carotinoid pigment, fucoxanthin, is found in two unusual binucleate dinoflagellates, *Durinskia baltica* and *Kryptoperidinium foliaceum,* which harbor a photosynthetic eucaryotic endosymbiont (Tomas and Cox 1973). The symbiont was determined to be a pennate diatom (Chesnick et al. 1997). A centric diatom endosymbiont, with nucleus, has been found in *Peridiniopsis penardii* (Takano et al. 2008). These species were widely studied (Tomas et al. 1973, Tippit and Pickett-Heaps 1976, Chesnick et al. 1996) as an intermediate stage between a colorless phagotrophic cell and a photosynthetic one (Pienaar 1980).

Blue-green dinoflagellates (see CP4) were hypothesized to harbor an endosymbiotic cryptomonad based on the presence of thylakoids in twos and nucleomorphs (Wilcox and Wedemayer 1984). Cryptomonads were ingested by colorless dinoflagellate cells which then showed epifluorescence, indicating chlorophyll and cryptophycean phycocyanin; pigmented cells had a second eucaryotic nucleus in the hypocone; plastids and nucleus were digested in about 10–14 days (Fields and Rhodes 1991). The absorption spectrum of pigments was the same for a blue-green *Gymnodinium* and a blue-green *Chroomonas* (Barsanti et al. 2009).

While many of the freshwater dinoflagellates are photosynthetic, others are heterotrophic or mixotrophic. Heterotrophy may take the form of uptake of dissolved material (*osmotrophy*), ingestion of whole prey (*phagotrophy*), production of a *peduncle* (feeding tube) from the sulcal area (*myzocytosis*) that sucks in prey contents, or formation of a *pseudopod* (feeding veil) (Jacobson and Anderson 1986). Prey include protozoans and metazoans, and phytoplankton including diatoms, dinoflagellates, and other flagellates (Table 1). Parasitic dinoflagellates pierce/digest a hole in the wall of filamentous green algae (*Oedogonium, Spirogyra, Zygnema*) and suck out the contents (Pfiester and Popovský 1979).

Mixotrophy is a strategy of combining photosynthesis with heterotrophy. Photosynthesis is supplemented by acquisition of dissolved inorganic nutrients, bacteria, or whole prey, or heterotrophy is supplemented by photosynthesis,

Table 1. Phagotrophy in freshwater dinoflagellates

Dinoflagellate	Prey	Reference
Esoptrodinium gemma	*Chlamydomonas*	Calado et al. 2006
Gymnodinium acidotum	*Chroomonas*	Fields and Rhodes 1991
G. austriacum	*Sphaerocystis lacustris*	Popovský 1985
G. helveticum	diatoms	Popovský 1982
	Gloeocystis, diatoms, *Eudorina*	Irish 1979
Katodinium spirodinoides	*Cyclotella, Aulocosira*	Popovský 1982
Prosoaulax lacustris	cryptomonads, chlorelloids	Calado et al. 1998
Tyrannodinium berolinense	*Phacus,* chlorophytes	Wedemayer and Wilcox 1984

and mixotrophy is found in both thecate and athecate taxa. One model for mixotrophic algae suggests that primarily photosynthetic organisms may take in nutrients from other sources when inorganic nutrients such as nitrogen and phosphorus are low, when light is limiting, or when trace factors are required (Stoecker 1998).

The cell covering in dinoflagellates has historically been described as armored/thecate or naked/unarmored/athecate. Armored dinoflagellates have a cellulose wall divided into plates; naked dinoflagellates lack such a wall. Thin sections examined by transmission electron microscopy (TEM) made this distinction less clear. Loeblich III (1969) resurrected the term *amphiesma* for the assimilative cell covering; it included an outer continuous membrane, a layer of vesicles, and, in armored species, a pellicular layer (Loeblich III 1969). The vesicles may be empty or contain thin platelike material in naked taxa and thick cellulosic plates in armored taxa. Loeblich III (1969) considered the plasma membrane to be interior to the amphiesma, although others consider the outer membrane to be the plasmalemma (Hansen et al. 2000) (Fig. 2).

Dinoflagellates have a large nucleus with permanently condensed chromosomes. This unique feature, sometimes called a *dinokaryon*, can be seen in the light microscope because the nucleus is so large and the condensed chromosomes look like the whorls of a fingerprint (see CP5[1]). The nucleus lacks histones and has a complex mechanism for chromosome duplication and separation (Dodge 1963, Dodge and Vickerman 1980). The shape and placement of the nucleus in the upper, middle, or lower section of the cell is part of the species description.

Figure 2. Thecal plates with an outer membrane on lower half of *Peridinium volzii*.

Many dinoflagellates have a red eyespot in the sulcus (see CP6[6]) and dinoflagellates are unusual in having at least five different types of eyespot (Dodge 1969). The most common occurs in dinoflagellates with peridinin-containing chloroplasts and consists of closely packed hexagonal lipid globules associated with thylakoids. Another type of eyespot is found in dinoflagellates harboring a photosynthetic endosymbiont where the eyespot is membrane bound but not associated with a chloroplast (Kawai and Kreimer 2000). Other types are a single row of small lipid globules in a chloroplast associated with bricklike structures (seen in TEM) (*Borghiella*, Moestrup et al. 2008), globules not associated with a membrane (*Tovellia coronata*, Lindberg et al. 2005), and a complex eyelike ocellus (*Warnowia*, Dodge 1969).

In addition to observable features, dinoflagellates have some unique internal features such as an osmoregulatory *pusule* (Dodge 1972) and ejectile organelles called *trichocysts*.

Morphology

By convention, an illustration of a dinoflagellate cell is considered to be facing the reader. The face with the sulcus is termed *ventral,* the back *dorsal,* the top half of the cell is the *epitheca/epicone,* the bottom half is the *hypotheca/hypocone* (Figs. 3, 4). The dividing groove between the epitheca and hypotheca is the *cingulum,* in which is located the *transverse flagellum.* The *longitudinal flagellum* lies in a groove on the ventral face termed the *sulcus.* If the epitheca is viewed from above, looking down toward the cingulum, the view is termed *apical* (Fig. 5). An indentation due to the sulcus identifies the ventral direction. Similarly, if the hypotheca is viewed from below, looking up at the cingulum, the view is termed *antapical* (Fig. 6). Many dinoflagellates show dorsoventral compression, from slight to almost the thickness of a potato chip in *Naiadinium biscutelliforme;* this is indicated in apical and antapical views as bean-shaped cross sections. One should remember that the viewer's right is the cell's left.

Dinoflagellates have traditionally been divided into armored/thecate and naked/unarmored/athecate groups based on the contents of vesicles beneath the plasma membrane. Naked dinoflagellates lack cellulose plates in the vesicles. Armored dinoflagellates have plates in the vesicles. The plates were named according to their shape or location. Kofoid (1909) proposed a numbering system for the plates based on his observation that the plates are arranged in approximately concentric series. The plates surrounding the apex he termed *apical* and numbered counterclockwise beginning with the most ventral. Apical plates are designated by a single prime marking after the plate, as in 1′ (Figs. 3, 6). Plates above the cingulum, the precingular plates, are designated by two prime (1″) and defined by Balech (1980) as those which border the cingulum but do not touch the pore plate (Fig. 6). Postcingular plates are

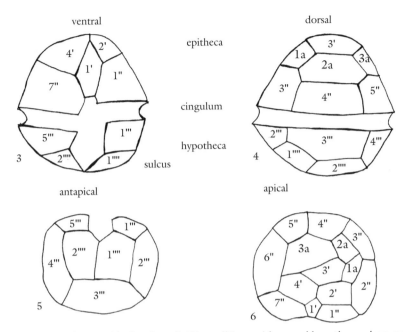

Figure 3. Ventral view with cingulum dividing cell into epitheca and hypotheca, plates numbered, sulcus present.

Figure 4. Dorsal view with cingulum dividing cell, plates numbered.

Figure 5. Antapical view, hypotheca, with postcingular (''') and antapical plates ('''') numbered.

Figure 6. Apical view, epitheca, apical ('), precingular ("), and apical intercalary (a) plates numbered.

three prime (1''') (Fig. 5). Antapical plates are four prime (1''''), in contact with the sulcus but not the cingulum (Balech 1980) (Figs. 3, 5). Cingular plates have a C preceding their number (Fig. 7). Sulcal plates are designated S (Fig. 8). Plates located between the apical and precingular series in the epitheca are designated "a" for anterior intercalary (Figs 4, 6), and plates in the hypotheca touching neither sulcus nor cingulum are designated "p" for posterior intercalary. A plate tabulation pattern is considered diagnostic for a taxon and is included in descriptions (e.g. 4', 3a, 7", C5, S5, 5''', 2'''' for *Peridinium cinctum*, meaning 4 apical plates, 3 apical intercalary, 7 precingular, 5 cingular, 5 sulcal, 5 postcingular, and 2 antapical).

Balech (1974) stressed the importance of the number and shape of the cingular and sulcal plates. He removed species in *Peridinium* into the genera *Protoperidinium, Scrippsiella,* and *Ensiculifera* based in part on cingular

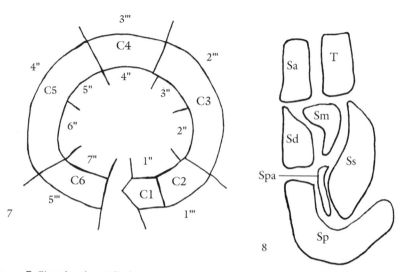

Figure 7. Cingular plates (C) showing relationships to pre- and postcingular plates.

Figure 8. Sulcal plates: T = transition, Sa = sulcal anterior, Sp = sulcal posterior, Ss = left sulcal, Sd = right sulcal, Sm = medial sulcal, Spa auxiliary posterior sulcal.

configuration. Boltovskoy (1999) removed *Glochidinium* from *Peridiniopsis* based in part on cingular plates. The first cingular plate may extend into the sulcus (left side); if small, it is called a *transition plate* (T). A plate on the right side, originating in the cingulum and terminating in the sulcus, is a *sulcal plate*.

Sulcal plates may be thinner than other plates in armored dinoflagellates and have been omitted in many illustrations. A terminology based on Balech (1974, 1980) is used herein (Fig. 8). The anterior plate is Sa (sulcal anterior); it may extend into the epitheca. The posterior plate is Sp. Both Sa and Sp may have ornamentation like that of other plates. The right sulcal plate (Sd, *dexter* = right) and left (Ss, *sinister* = left) flank the flagellar pore(s). Median (Sm) and accessory plates (Spa) are indicated as needed.

The apical pore present in some thecate dinoflagellates is surrounded by a pore plate (pp). The pore may be covered by a cover plate. A narrow plate on the ventral face of the cell between the pore plate and the 1′ plate is termed the *canal plate* (cp) (Dodge and Hermes 1981) (Fig. 9). Plates associated with an apical pore are included in the plate formula. The presence of an apical pore is frequently signaled by an apex that is apiculate (slightly drawn out), chimney-like, or drawn out into a horn (Fig. 10). An apical pore may also appear as a slight depression.

Morphological vocabulary for plate shapes and variability is extensive. The 1a plate may be ortho, meta, or para, denoting four, five, or six sides,

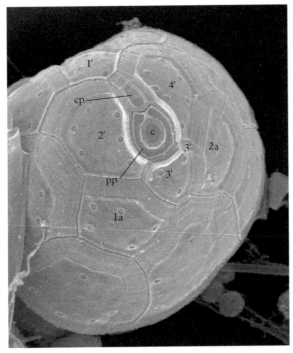

Figure 9. Plates associated with the apical pore: c = cover plate, pp = pore plate, cp = canal plate.

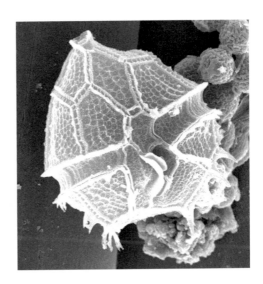

Figure 10. Apiculate apex indicating presence of an apical pore.

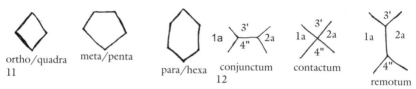

Figure 11. Shapes of the 1a and 2a plates.
Figure 12. Relationships between the 3' and 4" plates.

respectively (Fig. 11). For the 2a plate, the terms are quadra, penta, or hexa, respectively, to specify number of sides. The sutures between plates can shift, changing the shapes and contacts of plates; sutures have been designated (α between 1" and 2", β, γ, δ, ϵ, ψ). The relationship between the 3' and 4" in thecate dinoflagellates with anterior intercalary plates has been called *conjunctum* when they share a side, *contactum* when they just touch, and *remotum* when they are separated by the intercalary plates (Fig. 12) (Lefèvre 1932). Plates may also fuse, a *simplex* event, or divide, a *complexum* event (Lefèvre 1932).

Thecate dinoflagellates may have ornamentation on plates, particularly if the cells are older, that may be useful for identifications. *Reticulate* ornamentation is fairly common and consists of polygons with raised edges surrounding a trichocyst pore. Edges are less remarkable in young cells and may be exaggerated into small spines in some species. Other types of ornamentation on plates include granules and spines (Dodge 1983). Plate margins, particularly those near the apex, the cingulum, and sometimes the sulcus and antapex, may be extended into lists or wings.

Kofoid's system has been widely used since it was introduced (1909), but other systems have been devised that use a numeric or alphanumeric system to show patterns of relationships, particularly the sulcal plates as continuations of other series (Eaton 1980). Admittedly, the Kofoid system is artificial, but it remains a convenient and widely used method.

The taxonomy of dinoflagellates lacking plates is based on overall shape, size, cingulum position, features of the sulcus, and internal features. Chloroplast presence, number, color, and shape; presence and shape of an eyespot; and location of the nucleus are features visible in the light microscope that are used to identify naked dinoflagellates.

Some dinoflagellates seem to lack the defining dinoflagellate characteristics of golden color and cingulum. They are known to be dinoflagellates based on detailed work on life histories that always include a motile, cingulum-bearing life stage.

Ecology, Blooms

It is not my purpose in this work to provide extensive details on the ecology of freshwater dinoflagellates, and I direct the reader to Pollinger (1987) or

Carty (2003) for those; however, ecological aspects that facilitate identification of species will be reviewed briefly here.

A *bloom* is a huge preponderance of a single species in the phytoplankton. The water sample may be visibly colored by the blooming species, and a microscopic investigation of a water sample will confirm the bloom. The freshwater dinoflagellate species most often causing a bloom is *Ceratium hirundinella,* but other species are capable (Table 2). Blooms develop if growth conditions (light, temperature, nutrients) are optimal and ameliorating factors (predation, disease, wind) are minimal. Blooms often end with depletion of nutrients and formation of cysts. When a large bloom collapses, a fish kill may result (Nicholls et al. 1980). *Red tide,* a discoloration of the water due to a bloom, occasionally occurs in freshwater (Horne et al. 1972, Herrgesell et al. 1976), but reports of toxicity are rare.

Table 2. Some dinoflagellate species causing blooms in North America

Species	Location	Reference	Note
Gymnodinium acidotum	LA	Farmer and Roberts 1990	
	MO	Fields and Rhodes 1991	
Ceratium furcoides	NY	Carty (herein)	
Ceratium hirundinella	ON	Nicholls et al. 1980	
	NWT	Moore 1981	
	WI	Klemer and Barko 1991	
	UT	Whiting et al. 1978	
Dinosphaera palustris	MA	Prescott and Croasdale 1937	"discolor water"
Kansodinium ambiguum	TX	Carty 1986	
Palatinum apiculatus var. laevis	KS	Thompson 1950	March
Parvodinium deflandrei	MS	Canion and Ochs 2005	90% of population
Parvodinium inconspicuum	WV	Perez et al. 1994	99% total biovolume
	PA	Koryak 1978	95% total counts, acidic water
Parvodinium pusillum	GA	Stoneburner and Smock 1980	88% of cells, acidic
Peridiniopsis borgei	NC	Whitford and Schumacher 1984	
Peridiniopsis penardii	CA	Horne et al. 1972, Herrgesell et al. 1976	60% volume
Peridiniopsis polonicum	OK	Nolen et al. 1989	
Peridinium limbatum	ON	Yan and Stokes 1978	
	WI	Kim et al. 2004	
Peridinium willei	UT	Stewart and Blinn 1976	1.5×10^5 cells/L
Thompsodinium intermedium	WI	Matthew Harp (pers. comm.)	>10,000 cells/mL
	TX	Gilpin, pers.comm., 2009	4.7×10^4 cells/mL
	TX	Gilpin 2012	
	Belize	Carty (herein)	
Woloszynskia reticulata	OK	Pfiester et al. 1980	

Toxins

Peridiniopsis polonicum has been linked to an ichthyotoxic alkaloid called *glenodinine* that caused fish mortality during a bloom in a lake (Hashimoto et al. 1968). Other toxins, called *polonicum toxins*, have been isolated and were ichthyotoxic but showed low mouse lethality and no mutagenic effects (Oshima et al. 1989). In addition to effects on wild fish populations, *P. polonicum* has been shown to cause mortality at a fish farm (Roset et al. 2002).

There are a few reports of other potentially toxic species. *Peridinium bipes* was shown to kill cells of the cyanobacterium *Microcystis aeruginosa* (Wu et al. 1998). *Peridinium aciculiferum* produced a toxic material thought to be allelopathic in nature (Rengefors and Legrand 2001). *Amphidinium,* a marine genus, does have some species known to produce ichthyotoxins including amphidinolides (with antitumor activity) and amphidinoles (antifungal) (Murray et al. 2004). Freshwater species of *Amphidinium* have been transferred to the genus *Prosoaulax.*

Habitats

Specific information on the ecological requirements of dinoflagellate species, such that a particular species could be considered an indicator of a specific environment, is minimal. In small ponds I usually find nonmotile and attached species; in large lakes and reservoirs, *Ceratium* and *Naiadinium biscutelliforme;* in manmade water features treated with chemicals to minimize plant growth, heterotrophic species. The literature includes papers on acid-tolerant species (*Parvodinium inconspicuum*) in connection with acid rain and acid mine drainage (Koryak 1978, though *P. africanum* is illustrated), and other acid environments (Holopainen 1992, Fott et al. 1999). I consider *P. inconspicuum* acid tolerant as I find it everywhere. More likely candidates for acidophiles are species found in brown-water habitats (*Ceratium carolinianum, Peridinium limbatum*). Cold-preferring species (cryophiles) may be found in more northern areas like Greenland, Canada, and the northern United States, or during the winter in other places. *Gymnodinium pascheri* was found in a ditch and causing red snow in Ontario while temperatures remained below 10 °C (Gerrath and Nicholls 1974). *Peridinium aciculiferum* and *P. wisconsinense* have been found under the ice in North Dakota (Phillips and Fawley 2002), and *Borghiella tenuissima* can be a winter-blooming species (Crawford et al. 1970). *Peridinium aciculiferum* and *Woloszynskia ordinatum* are known to form large populations under the ice in Swedish lakes (Rengefors 1998). Care must be taken during examination of samples collected in the cold as cells may immediately ecdyse or disintegrate at room temperature.

Reproduction

Dinoflagellates have a life cycle that includes both asexual and sexual reproduction keyed to environmental conditions. Asexual reproduction occurs when a single cell divides to yield two. There are four different types of division based on the outer covering of the cell: (1) *Binary fission* occurs in athecate species with a parent cell splitting into two, the outer covering stretching and growing to cover daughter cells. Under the microscope, such dividing cells appear lumpy and without distinguishing characteristics. (2) *Desmoschisis* is seen in *Ceratium*, where a thecate parent cell divides along predetermined sutures and each daughter cell gets half the parental theca and generates the missing half (Fig. 13). In large populations of *Ceratium*, division may be observed, and strange-looking "half cells" and some "deformed" cells may be daughter cells completing regrowth (Fig. 14). In both these cases the plane of division is oblique, from upper right to lower left, and cells may remain motile while dividing. (3) *Eleutheroschisis* occurs when the outer thecate covering is shed and not used by the two daughter cells. The theca may be shed before or after cell division (Fensome et al. 1993). (4) A nonmotile resting cyst can be formed from an assimilative cell and may or may not have the characteristics of the assimilative cell. After a period of quiescence the cyst wall may dissolve or break apart, releasing one or two individuals.

Sexual reproduction has been documented in many freshwater dinoflagellates and is important for understanding the various forms and their complex life histories. Dinoflagellates are considered haploid, so a mitotic division can produce either two assimilative cells or two gametes that can be smaller than

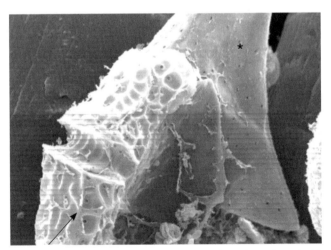

Figure 13. *Ceratium rhomvoides* with thick parental plates (*arrow*), and thin new plates (*asterisk*).

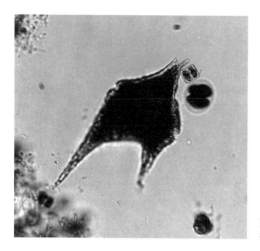

Figure 14. *Ceratium* showing growth of regenerating half cell.

the assimilative cells and may revert to them if not successful. Gametes unite in various positions, although usually sulcus to sulcus. In *Ceratium* a large female gamete swings open its sulcal plates to engulf a smaller male (Hickel 1988). Gametes in *Durinskia baltica* attach apex of one to sulcus of the other, forming a figure eight–shaped pair (Chesnick and Cox 1989). In *Peridinium inconspicuum*, gametes contact at the sulcus, then the protoplast from each moves out of its theca as plasmogamy, then karyogamy, occurs and a zygote forms (Pfiester et al. 1984). *Peridinium cinctum, P. willei,* and *P. gatunense* are described as having gametes that fuse laterally (Pfiester 1977). The resulting motile zygote has two trailing flagella (a *planozygote*) and becomes large and dark. Eventually motility is lost and the resulting *hypnozygote* can be recognized by its wide sutures (Fig. 15) (Pfiester 1984).

In planktonic, athecate, nonmotile *Cystodinium,* the parent cell releases two motile thecate zoospores that can change into the typical shape by ecdysing and developing a new cell wall, or can act as gametes and fuse apex of one cell to antapex of the other while motile. The resulting zygote is initially round and has a thick wall (Pfiester and Lynch 1980). The parasitic genera *Cystodinedria* and *Stylodinium* undergo reproduction following a feeding. The *Stylodinium* protoplast divides to produce two orange amoebae that when liberated crawl to new cells to feed, then revert to the *Stylodinium* shape. Alternatively, two motile gymnodinioid zoospores may be released that act as gametes (Pfiester and Popovský 1979).

METHODS

Field Sampling

Any freshwater habitat can be sampled for dinoflagellates, especially standing water such as ponds, lakes, and reservoirs. Other locations include streams,

Figure 15. *Peridinium gatunense* with wide suture bands (*asterisks*).

rivers, bogs, swamps, golf course water hazards, ice cores, farm ponds/stock tanks, holding ponds, fish hatchery ponds, home garden water features, moist sand, and drippy cliff faces. In addition to being in the water itself, dinoflagellates may be found attached to vegetation (especially filamentous green algae) and animals; submerged vegetation can be squeezed, and a turkey baster or plastic pipette can be used to collect surface sediments.

Samples may be collected as whole water and settled or centrifuged, or by using plankton nets (I prefer 10 μm because small dinoflagellates such as *Parvodinium inconspicuum* will pass through a 35 μm net) to concentrate plankton (although there is concern about concentrating zooplankton that eat the phytoplankton). Nets are useful in sparse algae conditions such as large lakes and reservoirs. Dinoflagellates are diurnal, migrating toward the surface during the day but avoiding the topmost layer, so sampling the entire water column is necessary.

Dinoflagellates are most common in temperate areas during the summer and will be found in many samples; however, samples taken throughout the year may yield cells, and seasonal sampling allows one to see the stages in the life history (dividing cells, assimilative cells, cysts). Some species are known to be cryophilic (cold loving).

Light Microscopy

It is a good idea to examine samples as soon as possible. Dinoflagellates are easily recognized in the living condition by (1) cingulum, (2) golden color, (3) and swimming pattern. Most dinoflagellates have at least one of these characteristics, and many have all three. Making sketches and noting general shape, features of the sulcus and cingulum, color, presence and location of spines, shape and location of plastids and accumulation bodies, presence of eyespot, presence of an apical pore, and size will aid in species identification. It may be difficult to tell thinly thecate species from naked species; thecate cells are more likely to have thicker, more defined cingular margins. As the microscope slide heats up and dries out, cells are stressed and may ecdyse if thecate, the protoplasm abandoning the theca, which can then be examined for plate pattern. Another method used to dissociate plates is introducing 6% sodium hypochlorite (chlorine bleach) under the coverslip (Boltovskoy 1975). Nonthecate dinoflagellates form a ball when stressed and no features can be seen, so early sketches may be critical.

One cautionary note: light microscopes reverse and invert the image, so while I sketch what I see, I try to remember that the cingulum is actually the mirror image of what I have sketched.

Scanning Electron Microscopy

Scanning electron microscopy (SEM) provides high-resolution images with details of the theca, sulcus, and plate patterns that may be difficult to see in the light microscope. Preserved samples are dehydrated in a graded alcohol series, critical point dried, coated with gold-palladium, then viewed in the microscope. Dinoflagellates have an outer membrane covering the plates that may obscure sutures (Fig. 2); there are techniques for killing cells that remove this membrane. Delicate nonthecate dinoflagellates may be observed following fixation in 1:1 2% osmium tetroxide, attachment to poly-L-lysine coated coverslips, and alcohol dehydrated followed by hexamethydisilazane (HMDS) (Botes et al. 2002). Other papers to consult about methods for delicate dinoflagellates include Steidinger et al. (1996b) and Truby (1997).

Preservation

The method of preservation depends on the intended future use of samples. Lugol's solution (5 g iodine, 10 g potassium iodide, 100 ml distilled water) is more likely to preserve flagella and is satisfactory if samples are going to be used in SEM. A little glycerol may be added to prevent total loss of the sample in case of dehydration. Since iodine stains the stored starch brown/black, it may be difficult to see plate patterns in the light microscope. If Lugol's is prepared without glycerol, cells remain useful for DNA analysis. Alternatively, glutaraldehyde does not discolor samples, and formaldehyde has also been used (0.7–5%).

Culturing

Many dinoflagellates are not amenable to culturing, and success requires patience. Single cells can be removed from a raw sample using a drawn pipette and serially rinsed before inoculation into a medium. A drawn pipette is made by heating the middle of the narrow end of a glass Pasteur pipette until it is soft, removing from heat, and pulling. Some media used for dinoflagellates include that of Carefoot (1968) and Medium Ch (Bruno and McLaughlin 1977). If culturing is a goal, then water from the locality is collected, filter sterilized, and used for dilutions and medium. It may be supplemented with defined growth medium. Fields and Rhodes (1991) were able to maintain phagotrophic *Gymnodinium acidotum* in culture by including its prey *Chroomonas*.

Carefoot media (Carefoot 1968)

$NaNO_3$	247 mg
$CaCl_2 \cdot 2H_2O$	11 mg
$MgSO_4 \cdot 7H_2O$	47 mg
K_2HPO_4	9 mg
KH_2PO_4	23 mg
NaCl	15 mg
PIV metals	5 mL
DW	995 mL
pH	7.5

PIV metals

$FeCl_3 \cdot 6H_2O$	19.6 mg
$MnCl_2 \cdot 4H_2O$	3.6 mg
$ZnSO_4 \cdot 7H_2O$	2.2 mg
$CoCl_2 \cdot 6H_2O$	0.4 mg
$Na_2MoO_4 \cdot 2H_2O$	0.25 mg
$Na_2EDTA \cdot 2H_2O$	100 mg
DW	100 mL

Medium Ch (Bruno and McLaughlin 1977)

$CaCl_2$	50 mg
$MgSO_4 \cdot 7H_2O$	100 mg
KCl	5 mg
$NaNO_3$	100 mg
K_2HPO_4	10 mg
PII metals mix	10 mL
Vitamin B_{12}	1 µg
Vitamin mix 22	2 mL
TRIS	200 mg
DW to make	1000 mL
pH	7.0–7.2

PII metals mix/L: FeCl 10 mg, ZnCl 5 mg, H3BO3 200 mg, MnCl 40 mg, CoCl 1 mg, Na_2EDTA 1 g.

Vitamin mix 22/L: thiamine · HCl 200 mg, nicotinic acid 100 mg, biotin 500 µg, B_{12} 50 µg.

TAXONOMY

Evolutionary relationships among organisms are better understood if the organisms are separated into clearly delineated taxa and patterns of differences can be described. For dinoflagellates the patterns have traditionally been based on external morphology and internal anatomy. Habitat considerations (i.e., freshwater vs. marine, planktonic, epizoic, epiphytic) have become vital for identifying genera. For thecate dinoflagellates, the patterns have centered on the arrangement of plates. Widespread patterns, such as a hypotheca of five postcingular and two antapical plates (all species of *Peridinium, Protoperidinium, Peridiniopsis, Ceratium, Durinskia*), suggest common ancestry in the taxa that display them and focus attention on differences in the epithecal, cingular, and sulcal plates. Other widespread patterns such as the presence of an apical pore and peridinin-containing chloroplasts suggest that species without them are derived. Reconstructing ancestral patterns, then hypothesizing fusion, splitting, and shifting events to give rise to derivative patterns can provide some insight into the evolutionary history of the group.

Previous emphasis on morphology for recognizing relationships among dinoflagellates has given way to a polyphasic species definition that includes molecular phylogenetic data, external morphology, internal anatomy, life cycle features, and ecology. It is important that all aspects of the species be considered when drawing phylogenetic hypotheses. Current molecular analysis has focused strongly on large-subunit (LSU) and small-subunit (SSU) rDNA, with only a few protein-coding genes investigated. It would be useful to investigate other dinoflagellate genes.

Currently the primary reference for hierarchy in dinoflagellates is Fensome et al. (1993), *A classification of living and fossil dinoflagellates*. Research since its publication and the profusion of "Family Uncertain" designations make that scheme no longer useful. At this time, higher levels of classification above genus are too uncertain; thus none are provided here.

Certain genera require more discussion than is provided in the principally descriptive paragraphs of the genus and species that follow. The taxonomic history and problems of seven genera are reviewed here.

Glenodinium Ehrenberg 1835

Glenodinium was erected by Ehrenberg for organisms having an eyespot, to distinguish them from those in the genus *Peridinium* that did not have an eyespot. Eyespot presence proved less consistent than other features of the genera. An examination of synonymies shows that many thecate species were assigned to *Peridinium* or *Glenodinium* at one time (Table 3).

Ehrenberg's 1838 illustrations show cells with a fringe of cilia at the cingulum, and the apex down, but they are clearly dinoflagellates. The illustrations

Table 3. Changes to species in *Glenodinium*

Former name	Current name
Glenodinium acutum	*Entzia acuta*
Glenodinium ambiguum	*Kansodinium ambiguum*
Glenodinium balticum	*Durinskia dybowski*
Glenodinium berolinense	*Tyrannodinium berolinense*
Glenodinium dinobryonis	*Staszicella dinobryonis*
Glenodinium neglectum	*Jadwigia applanata*
Glenodinium palustre	*Dinosphaera palustra*
Glenodinium penardiforme	*Glochidinium penardiforme*
Glenodinium polylophum	*Lophodinium polylophum*

of *Glenodinium tabulatum* show at least two taxa whose outlines are roughly angular-pentagonal; cells a and *b* are probably the same, with an oval eyespot in the sulcus, an apical pore, and green chromatophores. Cell *c* has many plates in the epitheca and hypotheca, and *d* is thecate and dorsoventrally compressed (Ehrenberg 1838). *Glenodinium apiculatum* illustrations show cells with an oval eyespot in the sulcus, broad sutures between plates, and cilia-like fringes at plate margins. It is possible the noncingular cilia represent lists or the raised growth margins of plates. The cells seem to have fewer plates than *G. tabulatum* and are spherical-elliptical in outline (Ehrenberg 1838). *Glenodinium cinctum*, type for the genus, is illustrated showing cells with no indication of plates, although cingulum and sulcus are evident. The cells are elliptical to spherical, with oval shapes also present. They contain tan chromatophores, and a distinctive, red, horseshoe-shaped eyespot. The size given is approximately "1/48 of a Paris line" (50 μm) (Ehrenberg 1838).

An early citation of *Glenodinium cinctum* was made by Stein (1883), who illustrated and described it as containing light brown chromatophores, a longitudinal sausage-shaped nucleus in the epitheca, red or bloodlike oil bodies (one U- or V-shaped) and a cingulum illustrated as having many cilia; there is no evidence of plates. The same paper includes illustrations of many thecate species with clear delineation of plates, including those of *Peridinium cinctum* that have been universally accepted as the plate pattern (Stein 1883). Levander (1892) illustrated a plate pattern of 4′, 2a, 6″, 5‴, 2⁗ for a cell he identified as *G. cinctum;* in 1894, however, he decided that the species illustrated in 1892 was not *G. cinctum*, named it *G. balticum*, and described it as having a gutter-shaped eyespot (Levander 1894). Bourrelly (1968a) and Loeblich (1980) accepted that the Levander 1892 illustration is not *G. cinctum.*

In 1916 Wołoszyńska established a new thecate genus *Sphaerodinium*, with a plate pattern of 4′, 4a, 7″, 6‴, 2⁗, and named three species. *Sphaerodinium polonicum* had a horseshoe-shaped eyespot; the other two species, *S. limneticum* and *S. cracoviense*, had eyespots of uncertain shapes (Wołoszyńska 1916). The following year she synonymized *Glenodinium cinctum* with *Sphaerodinium*, listing the new combination *Sphaerodinium cinctum* (Ehren.) Wołoszyńska. She stated that *S. polonicum* and *S. cracoviense* belonged with the new combination, and that *S. limneticum* was identical to *Glenodinium cinctum* (Wołoszyńska 1917). The differences that originally separated her three species such as overall shape, shape of the nucleus, shape or absence of the eyespot, were considered variations within *S. cinctum*. While only *S. polonicum* was originally described as having a horseshoe-shaped eyespot, subsequent authors have cited *S. cinctum* based on *S. cracoviense*, retaining *S. polonicum* as a separate species (Schiller 1937, Huber-Pestalozzi 1950, Starmach 1974). The first two authors retained the genus *Glenodinium*, though not *Glenodinium cinctum*. Bourrelly suggested that *Glenodinium* was poorly defined and advocated the transfer of the species of *Glenodinium*, defined by Schiller as having

3–5', 0–1a, 6–7", 5''', 2'''', to *Peridiniopsis* (Bourrelly 1968a). He listed two species of *Sphaerodinium* (*S. polonicum* and *S. fimbriatum*) (Bourrelly 1970). Starmach (1974) abandoned *Glenodinium* for *Peridiniopsis* Lemmermann.

Loeblich (1980) reviewed the complex problem of the type species of *Glenodinium*, and while concurring with Wołoszyńska that *Sphaerodinium* was congeneric, cited ICBN rules that the name should be *Glenodinium cinctum* with the plate tabulation formula of *Sphaerodinium*. He did not discuss the status of the other species of *Glenodinium* or *Sphaerodinium*.

At the time Ehrenberg illustrated *Glenodinium* and *Peridinium*, the difference between them was the presence of an eyespot in the former; interestingly, however, he illustrated the species of *Glenodinium* with plates but did not do so for *Peridinium*. The species of *Peridinium* illustrated by Ehrenberg have since been segregated into several genera, including *Gymnodinium*, *Peridinium*, *Ceratium*, and *Protoperidinium*. The practice of taxonomists has been to relegate thick-walled species to *Peridinium* and thin-walled species to *Glenodinium* (Prager 1963); there has been a vast exchange of species between the two genera. As plates became more apparent, species were segregated to *Glenodinium* with a pattern of 3–5', 0–1a, 6–7", 5''', 2'''' and to *Peridinium* with 4', 2–3a, 7", 5''', 2'''' (Schiller 1937). *Glenodinium* still contains species of unknown plate pattern, the cells illustrated and identified by general outline, size, and internal features.

Glenodinium cinctum (the type species defining the genus) was originally depicted without plates and with a horseshoe-shaped eyespot. When Stein (1883) illustrated the plate patterns of many thecate species, including *Parvodinium umbonatum*, which has thinner plates than *Peridinium*, no plates were shown for *G. cinctum*. The lack of an accepted plate pattern leaves the species based on the shape of an eyespot. An undefined type species leaves the genus undefined. Bourrelly's (1968a) proposal of transferring *Glenodinium* species to *Peridiniopsis* seems the most workable, but see the following discussion of *Peridiniopsis*.

Peridiniopsis Lemmermann 1904

The genus description states that the cingulum is distinct and twisted, the sulcus narrow and reaching the end, the 1' plate reaching the apex. Plates include 6 precingular, 5 postcingular, and 2 antapical. The illustration of *Peridiniosis borgei* Lemm., the type species, accompanying the description shows a plate tabulation of 3', 1a, 6", 5''', 2'''', with an apical pore (Lemmermann 1904). Bourrelly (1968a, b), in reviewing the problem of *Glenodinium*, chose *Peridiniopsis* as a more defined substitute. He cited Schiller's description of 3–5', 0–1a, 6–7", 5''', 2'''' for the genus. He arranged species in categories based on epithecal arrangement, for example, 3'-1a-6", 4'-0a-6", 5'-0a-6", 3'-1a-7", 4'-0a-7", 4'-1a-7", 5'-0a-7", and 5'-1a-7" (Table 4). Starmach (1974) followed Bourrelly's groupings and listed 24 species of *Peridiniopsis*.

Table 4. Epithecal plate patterns for related thecate genera, and species of *Peridiniopsis*

Precingular plates	Apical plates		
	3'	4'	5'
5"	1a *Kansodinium*	0a *Gonyaulax*	
6"	1a *Dinosphaera*	0a *Periop. penardii*	0a *Periop. cunningtonii*
	1a *Glochidinium*	0a *Glochidinium*	
	1a *Periop. borgei, edax, kulczynski*	0a *Tyrannodinium*	
		1a *Periop. cunningtonii*	
		2a *Durinskia*	
		3a *Thompsodinium*	
7"	1a *Periop. lindemanii, oculatum*	0a *Periop. elpatiewskyi*	0a *Periop. thompsonii*
		1a *Staszicella*	1a *Periop. quadridens*
		1–2a *Entzia, Naiadinium*	
		2a *Parvodinium*	
		3a *Peridinium*	
		4a *Sphaerodinium*	

Peridiniopsis is an improvement over *Glenodinium*, but the wide range of plate patterns is not satisfactory in one genus. *Peridiniopsis* had become the grab-bag that was *Glenodinium*. Boltovskoy (1999) began the removal of taxa from *Peridinopsis* by erecting *Glochidinium* (3'-1a or 4'-0a, 6", C3, S4, 5''', 2'''') for *Peridinium/Peridiniopsis penardiforme*, based in part on a different number of cingular plates from that of the type species. Next was *Tyrannodinium* (4', 0a, 6", C6, 5''', 2''''), a heterotrophic genus lacking an eyespot and characterized, in part, by its feeding behavior. *Tyrannodinium* was based on *Peridinium/Peridiniopsis berolinense* (Calado et al. 2009). Appendix C of this volume proposes *Naiadinium* for *Peridinium/Peridiniopsis polonicum* based on morphological and molecular analyses. Ultrastructural data and molecular sequences will undoubtedly support the division of *Peridiniopsis* into more separate genera (Logares et al. 2007a).

Peridinium Ehrenberg 1830

Ehrenberg erected the genus *Peridinium* in 1830. Later, he described the genus as "Animal de la famille de Péridinés, ayant de la carapace (membraneuse) un sillion transveral cilié et point d'oeil" (Ehrenberg 1838, 252). *Peridinium* was distinguished from *Glenodinium* by the presence of an eyespot in the latter. (Note: while no eyespot is visible in the light microscope, TEM work has shown one near the longitudinal microtubular root in *Peridinium* [Calado et al. 1999]). Ehrenberg recognized the similarity to dinoflagellates of various species described in animal genera and transferred them to either *Peridinium* or *Glenodinium*. The species have since been transferred to several dinoflagellate genera, including *Gymnodinium*, *Ceratium*, and *Protoperidinium*. The species of *Peridinium* illustrated in the 1838 atlas include *Peridinium cinctum* (O.F.M.) Ehrenberg, the type species, a freshwater species.

Table 5. Groups in *Peridinium*

Group	Distinguishing features	Species
Cinctum	no apical pore, 1' remote from 3', 4''' penta	*cinctum, gatunense, raciborsii*
Willei	no apical pore, 1' remote from 3', 4''' quadra	*volzii, willei*
Striolatum	no apical pore, 1' touches 3'	*striolatum*
Bipes	apical pore, 4''' contacts 2a	*achromaticum, bipes, limbatum*
Lomnicki	apical pore, 4''' contacts 2a, 3a, 1a penta	*aciculiferum, godlewski, keyense, lomnickii, wierzejskii,*
Allorgei	apical pore, 4''' contacts 2a, 3a, 1a quadra	*allorgei, wisconsinsense*
Gutwinskii	apical pore, 4''' contacts 1a, 2a, 3a	*gutwinskii*
Subsalsum	apical pore, 4''' contacts 2a, 3', 3a	*subsalsum*

The description of the genus was broad; as more information was gathered about dinoflagellates, genera were extracted. The works of Balach (1974, 1980) and Abé (1981) have shown the differences between marine species and the type species. The plate tabulation pattern for *Peridinium* is now 4', 3a, 7'', 5''', 2'''' since the type species, *P. cinctum*, is 3a. The inclusion of thecate species with 2a was a holdover from the early inclusion of all noticeably thecate species in *Peridinium*. The two groups of species with a 2a tabulation are now separate genera, *Palatinus* for those lacking an apical pore (Craveiro et al. 2009) and *Parvodinium* for those with an apical pore (Carty 2008). While the groups of species within *Peridiniopsis* differ in apical plate tabulation, all species of *Peridinium* are 4', 3a, 7'', and the groups differ in the presence/absence of an apical pore and apical plate symmetry (Table 5). Molecular analysis of SSU rDNA of 238 sequences of marine and freshwater species showed strong grouping of *Peridinium willei* and *P. volzii*, which also grouped with freshwater species *P. bipes*, *P. gatunense*, and *P. cinctum*. *Peridinium wierzejskii* and *P. aciculiferum* (both in Group Lomnickii) were separated from the others and grouped with marine/estuarine species (Logares et al. 2007a). It is likely that species or species groups will be removed from *Peridinium*, possibly leaving only the type species group (Fensome et al. 1993).

Hemidinium–Gloeodinium, One Genus or Two?

Gloeodinium is considered by some (Kelley and Pfiester 1989) to be a stage in the life cycle of *Hemidinium*, as *Gloeodinium* produces *Hemidinium*-like zoospores and *Hemidinium* produces a *Gloeodinium*-like resting stage. However, fusion of these genera (Pfiester and Highfill 1993) may be premature. One study of vegetative reproduction of *G. montanum* found *Hemidinium nasutum*-like zoospores (Killian 1924), while another found *H. ochraceum*-like swarmers (Kelley and Pfiester 1989). Baumeister (1943) considered *G. montanum* to produce unequal-sized daughter cells while *H. nasutum* produced equal sized. Molecular work using SSU rDNA placed *Gloeodinium* and *Hemidinium* distant from each other (Logares et al. 2007a). Field experience and distribution maps of the genera further support that the two are found independently of each other and should continue to be recognized as separate genera.

Cystodinium–Cystodinedria–Dinococcus–Hypnodinium, Nonmotile, Athecate Genera

This quartet of immobile genera may all be one genus with the different generic names representing different forms or parts of one life cycle. *Cystodinium* is a planktonic, fusiform to lemon-shaped assimilative genus whose cells divide to produce mobile or immobile zoospores or gametes. The result of fusion of gametes produces a round *Hypnodinium*-like cell (Pfiester and Lynch 1980). *Cystodinium bataviense* has been recorded attached to, and parasitizing, *Oedogonium, Spirogyra,* and *Mougeotia,* much like the *Cystodinium*-shaped, but attached, *Cystodinedria* (Pfiester and Lynch 1980). *Cystodinium bataviense* has also been recorded as attached with a short stalk much like *Dinococcus* (Pfiester and Lynch 1980). Evidence for keeping the genera separate includes the distinctive rosettes of plastids and daisy-like arrangement of vacuoles in *Hypnodinium;* the smaller size and attached, obviously parasitic nutrition in *Cystodinedria;* and the smaller, attached cell with terminal spines in *Dinococcus.* As with *Hemidinium-Gloeodinium,* reports in the literature rarely list these genera occurring together, so I choose to retain all four generic names.

Amphidinium–Gymnodinium–Gyrodinium–Katodinium–Opisthoaulax–Prosoaulax, Motile, Athecate Genera

These athecate genera are separated based on cingulum position; supramedian in *Amphidinium* and *Prosoaulax;* median in *Gymnodinium;* offset in *Gyrodinium,* and submedian in *Katodinium* and *Opisthoaulax.* There are several considerations when choosing the genus to which a species belongs. Examination of species illustrations and synonymies shows that there is not always a consensus on cingulum position. Further, since most of the nonmotile taxa produce gymnodinioid-like motile cells, short lived zoospores or gametes might be mistaken for assimilative cells and be named species of *Gymnodinium.* A third concern is the decision by geologists that the motile form of a taxon be considered the correct name (versus the spore form name which might have priority), which would mean more species of *Gymnodinium* and fewer of *Cystodinium, Cystodinedria, Tetradinium, Dinococcus,* and others. Molecular analyses of SSU rRNA show that *Gymnodinium* species are not monophyletic (Saldarriaga et al. 2004), but this could be an artifact of incorrect names on GenBank sequences. Recent revision of *Amphidinium* limits this genus to only those species with a small, left-deflected epicone (Jørgensen et al. 2004). A new genus, *Prosoaulax,* was erected for freshwater species (Calado and Moestrup 2005). By this definition, *Amphidinium* is a mostly marine genus and *Prosoaulax* a heterotrophic one. Freshwater, photosynthetic species with small epicones and large hypocones, though not *Amphidinium* sensu stricto, will remain in that genus until moved.

It is often difficult to distinguish athecate from lightly thecate genera, and *Gymnodinium* is the most speciose genus in this volume. Even so, there remain many undescribed species in North America. Several species first identified as athecate *Gymnodinium* have been transferred to the lightly thecate *Borghiella*, *Tovellia*, and *Woloszynskia* (see next discussion). If there are enough cells in a sample, and they can be observed dying, athecate taxa form featureless round balls, whereas thecate taxa may contract away from the theca or ecdyse, in either case showing the presence of a theca. Recent emending of the description of *Gymnodinium* based on the type species, *G. fuscum*, now includes a horseshoe-shaped apical groove encircling the apex. This groove can occasionally be seen in the SEM, or surmised in the light microscope if the sulcus seems to extend deeply into the epitheca (see Figs. 28, 29). Most named *Gymnodinium* species have not been examined for the presence of this groove. Cingulum location has not been satisfactory in delimiting *Gymnodinium* from *Katodinium* or *Amphidinium,* and work is needed.

Woloszynskia–Jadwigia–Tovellia–Borghiella

In 1917 Wołoszyńska illustrated some species of *Gymnodinium* with numerous, thin, hexagonal to polygonal plates (also for *Glenodinium neglectum*). Thompson (1950), using iodine fumes to shrink away the protoplast, also saw plate sutures in thinly thecate species, though not in true species of *Gymnodinium*. He established the genus *Woloszynskia* for the species seen by Wołoszyńska, for species described by others but found to have numerous thin plates, and for a new species he found in Kansas, *Woloszynskia reticulata*. Improved cell fixation procedures and availability of SEM have uncovered other woloszynskioid species. The term *woloszynskioid* refers to the presence of a theca with numerous thin plates; other morphological features, including the cyst and apical plates indicated that the genus might be heterogeneous. Lindberg et al. (2005) examined several species of *Woloszynskia* and divided the genus into new genera based largely on morphology of apical plates (an apical line of narrow plates, ALP), LSU rDNA, cyst shape, and eyespot construction. *Tovellia* (for *W. coronata, W. coronata* var. *glabrum, W. apiculata, W. leopoliense, W. nygaardi,* and *W. stoschii*) has a lobed cyst with extended poles, ALP of two rows of small rectangular cells bordering a narrow band with punctae, and an unusual eyespot of pigment globules not associated with a chloroplast. *Jadwigia* (*W. neglecta*) has a round cyst, ALP of a single row of rectangular cells and a nonchloroplast eyespot (Lindberg et al. 2005). *Borghiella* (*W. tenuissima*) has a round, smooth cyst, ALP of a narrow band with punctae, and eyespot contained in a chloroplast (Moestrup et al. 2008). The type species, *W. reticulata*, remains in the genus *Woloszynskia*, as do species not specifically moved to other genera.

Keys and Taxa Descriptions

Taxa descriptions include the following:

- The citation for the name of a genus or species, the original reference, and synonyms used in reports from North America. Places I have collected are indicated by SC and my code (e.g., SC 830107-3 means the sample was collected in 1983, first month, seventh day, third sample); JH samples were collected by John D. Hall; VF by Victor Fazio.
- A diagnosis, with key distinguishing features. Following the diagnosis of a species is the number of the figure (Fig.), black-and-white plate (PL), and/or color plate (CP) herein that illustrates the species.
- A description of either the genus or the species. Species descriptions are categorized by external features, internal features, size, ecological notes, and location references (if fewer than 5 locations) or a map. Location references may include unpublished works and are found in Appendix A.

For species I have seen, the description is based largely on my own observations. For species I have not seen, descriptions come from the literature. Similarly, line drawings include original depictions, drawings from North American sources, and sketches of species I have seen.

Key to All Genera 1: Classic

1a. Cell with cingulum, photosynthetic (golden color), or both 2
1b. No reason to think it's a dinoflagellate (including parasites, look at diagrams)...
....46
2a. Cell golden but lacking a cingulum 3
2b. Cell with cingulum, golden or not.. 10
3a. Motile with apical flagella, thecate, oval cell dorsoventrally flattened ....
.... *Prorocentrum*

3b. Cell nonmotile 4
4a. Free floating, poles rounded or drawn into spines 5
4b. Cell attached 7
5a. Cell round 6
5b. Cell oval, with or without terminal spines .... *Cystodinium*
6a. Cell single .... *Hypnodinium*
6b. Cell(s) associated in layered mucilage .... *Gloeodinium*
7a. Attached to wet rocks/drip walls ....*Rufusiella*
7b. Attached to filamentous green algae 8
8a. Cell with spines.. 9
8b. Cell without spines... *Cystodinedria*
9a. Oval cell with polar spines, small attachment disk .... *Dinococcus*
9b. Tetragonal with spines at corners.. *Tetradinium*
9c. Irregular polygonal, corners with blunt spines .... *Dinastridium*
10a. Cingulum around only half the cell 11
10b. Cingulum encircles cell 13
11a. Cell golden, thin theca .... *Hemidinium*
11b. Cell nonphotosynthetic, naked 12
12a. Cingulum on left side.. *Bernardinium*
12b. Cingulum on right side .... *Esoptrodinium*
13a. Cingulum in upper third of cell 14
13b. Cingulum median, in lower third, or distinctly offset 16
14a. Cell with cellulose plates, sand dwelling................................ *Amphidiniopsis*
14b. Cell naked, planktonic.. 15
15a. Cell photosynthetic, golden or blue green ....*Amphidinium*
15b. Cell heterotrophic..*Prosoaulax*
16a. Cingulum in lower third of cell 17
16b. Cingulum median or distinctly offset.. 18
17a. Cell with or without plastids, no eyespot .... *Katodinium*
17b. Cell heterotrophic, with eyespot.. *Opisthoaulax*
18a. Cingulum distinctly offset, sulcus often at an angle 19
18b. Cingulum median 20
19a. Thecate cell.. *Gonyaulax*
19b. Athecate cell .... *Gyrodinium*
20a. Siliceous pentasters enclose nucleus, Arctic .... *Actiniscus*
20b. No siliceous pentasters 21
21a. Cell with 1 apical and 2–3 posterior horns...
....*Ceratium* (but see *Peridinium limbatum*)
21b. Cell without such horns 22
22a. Cell attached... 23
22b. Cell not attached.. 24
23a. Cell attached to sand grains in fast-moving streams........................*Dinamoeba*
23b. Attached to *Dinobryon* ....*Staszicella*
24a. Cell with longitudinal ridges .... *Lophodinium*
24b. Without such ridges 25
25a. Cell naked, without walls 26

47a. Forms gelatinous golden film on fish ....*Haidadinium*
47b. Forms yellowish dusting, parasitic ....*Piscinoodinium*
48a. Cell stalked .... *Stylodinium*
48b. Cell sessile... *Cystodinedria*

Key to All Genera 2: Starting with Habitat

1a. Collected free swimming or floating in water (pond, lake) 12
1b. Collected otherwise 2
2a. Associated with lithic substrate 3
2b. Associated with living substrate 5
3a. Collected from sand...4
3b. Collected from drip wall...*Rufusiella*
4a. Fast-running cold-water stream, photosynthetic, athecate, on pebbles.............
    ....*Dinamoeba*
4b. Edge of lake, nonphotosynthetic, thecate, in sand.................... *Amphidiniopsis*
5a. Epizoic 6
5b. Epiphytic 9
6a. Attached to fish...7
6b. Attached to zooplankton...8
7a. Yellow dusting on scales of minnows, guppies, and similar fish .. *Piscinoodinium*
7b. Gelatinous golden film on stickleback.......................................*Haidadinium*
8a. Tetrahedral with spines at four corners ....*Tetradinium*
8b. Irregular shape, spines at the 5–7 lobes.................................... *Dinastridium*
9a. Attached to *Dinobryon*..*Staszicella*
9b. Attached to filamentous algae (esp. *Oedogonium*) 10
10a. Cell with distinct stalk .... *Stylodinium*
10b. Cell sessile or with short stalk..11
11a. Cell sessile, oval.. *Cystodinedria*
11b. Cell with spines at poles ....*Dinococcus*
12a. Cell nonmotile..13
12b. Cell motile...15
13a. Cell oval, may have polar spines.. *Cystodinium*
13b. Cell spherical 14
14a. Cell singular in plankton .... *Hypnodinium*
14b. Cell in colonial mucilage with others, associated with *Sphagnum* .... *Gloeocystis*
15a. From high Arctic lakes, with siliceous pentaster .... *Actiniscus*
15b. Cell without pentasters...16
16a. Cell heterotrophic..17
16b. Cell photosynthetic..26
17a. Cell naked...18
17b. Cell thecate...22
18a. Cingulum incomplete..19
18b. Cingulum complete 20
19a. Right-descending cingulum.. *Bernardinium*
19b. Left-descending cingulum, red accumulation bodies................... *Esoptrodinium*
20a. Cingulum median 21

20b. Cingulum submedian .... *Opisthoaulax*

21a. Cell covering smooth .... *Gymnodinium* (in part)

21b. Cell with vertical rows of punctae, Arctic .... *Pseudoactiniscus*

22a. Posterior bilobed .... *Glochidinium*

22b. Posterior otherwise, single spine may be present 23

23a. Cell wider than long, extended apical pore, sulcal list .... *Entzia*

23b. Cell otherwise 24

24a. Obvious single slender spine at antapex .... *Parvodinium goslaviense*

24b. Antapex without obvious spine 25

25a. Apical pore present .... *Tyrannodinium*

25b. Apical pore absent .... *Peridiniopsis edax*

26a. No obvious cingulum, flagella apical .... *Prorocentrum*

26b. Cingulum incomplete .... *Hemidinium*

26c. Cingulum encircles cell 27

27a. Cell without plates 28

27b. Cell with plates 32

28a. Cingulum median .... *Gymnodinium*

28b. Cingulum above or below median or skewed 29

29a. Cingulum above median, small epicone 30

29b. Cingulum below median or skewed 31

30a. Cell photosynthetic, yellow gold or blue green .... *Amphidinium*

30b. Cell heterotrophic .... *Prosoaulax*

31a. Cingulum below median, hypocone small .... *Katodinium*

31b. Cingulum and sulcus skewed .... *Gyrodinium*

32a. Cell with longitudinal ridges .... *Lophodinium*

32b. Cell otherwise 33

33a. Plates obvious, either thin or thick 34

33b. Plates barely discernible 38

33c. Hypothecal plates thick, epithecal barely discernible ...... *Woloszynskia reticulata*

34a. Cingulum and sulcus look twisted, obvious apical pore .... *Gonyaulax*

34b. Cingulum and sulcus mostly perpendicular 35

35a. Cells small (11–25 μm diam.), may have spines .... *Parvodinium*

35b. Cells larger (30–55 μm diam.) 36

36a. Cell with antapical list, large postcingular plates .... *Thompsodinium*

36b. Pattern 4′, 2a, 7″, no apical pore .... *Palatinus*

36c. Cell otherwise 37

37a. Apical pattern 4′, 2–3a, 7″, may have apical pore, spines, lists .... *Peridinium*

37b. Apical pattern 3–5′, 0–1a, 6–7″, with apical pore, many species with spines ....
.... *Peridiniopsis*

38a. Cell without eyespot, without apical pore 39

38b. Cell with eyespot in sulcus 41

39a. Cell obovate, plates small, numerous .... *Woloszynskia cestocoetes*

39b. Cell round to oval, plates otherwise 40

40a. Cell round, no dorsoventral compression .... *Dinosphaera*

40b. Cell oval, dorsoventrally compressed .... *Glenodiniopsis*

41a. Three or four plates around apical pore 42

41b. Many polygonal plates 44

42a. One antapical plate...*Kansodinium*
42b. Two antapical plates 43
43a. Two apical intercalary plates, asymmetrical apical arrangement..........*Durinskia*
43b. Four antapical plates, symmetrical apical arrangement............... *Sphaerodinium*
44a. Strong dorsoventral compression, cryophiles ....*Borghiella*
44b. Cell not flattened 45
 45. Many small hexagonal plates.. *Jadwigia*
45b. Numerous polygonal plates ....*Tovellia*

Key to All Genera 3: Common or Easy-to-Identify Freshwater Species

Be sure to read the descriptions, as there are additional blue-green and color-less species as well as other species of *Ceratium* not mentioned in this key.

 1a. Cell with single, elongate, apical horn; 3 posterior horns...................................
    ....*Ceratium hirundinella*
 1b. Cell with single, short, apical horn; 2 posterior horns ..... *Peridinium limbatum*
 1c. Cell without horns, may have spines...2
 2a. Cell colorless, may have colored accumulation bodies...................................3
 2b. Cell with chloroplasts...4
 3a. Cell rhomboidal, sulcal list ....*Entzia acuta*
 3b. Cell with bilobed posterior................................... *Glochidinium penardiforme*
 4a. Cell blue green.. *Gymnodinium aeruginosum*
 4b. Cell golden 5
 5a. Bottom with heavy plates, top with thin plates.............. *Woloszynskia reticulata*
 5b. Bottom and top halves similar 6
 6a. Cell diamond shaped.. *Peridinium wisconsinense*
 6b. Cell otherwise..7
 7a. Cell wider than long; looks like Saturn with rings .... *Peridinium gatunense*
 7b. Cell otherwise..8
 8a. Cell ovoid, with fringes top and bottom, large 1′ plate...........*Peridinium willei*
 8b. Otherwise 9
 9a. Cells small, 10–25 µm diameter...10
 9b. Cells larger, 30 µm or more...11
10a. Cell pentagonal, may have spines..........................*Parvodinium inconspicuum*
10b. Cell ovoid, may have spines.....................................*Parvodinium umbonatum*
11a. Cell long, 50–80 µm, shaped like ice cream cone........... *Gymnodinium fuscum*
11b. Otherwise 12
12a. Cell with posterior fringe.....................................*Thompsodinium intermedium*
12b. Cell with posterior spine(s)..13
13a. Single stout spine .... *Peridiniopsis polonicum*
13b. Several spines, also on postcingular plates ....*Peridiniopsis quadridens*

ATHECATE/NAKED TAXA

Actiniscus

(Ehrenberg) Ehrenberg 1843, 100–106

BASIONYM: *Dictyocha* subgenus *Actiniscus*. Ehrenberg 1841, 147.

TYPE SPECIES: *Actiniscus pentasterias*. Ehrenberg 1843.

Mostly marine, planktonic, nonphotosynthetic, with siliceous internal pentasters.

Predominantly a marine genus (one freshwater species), well known from the fossil record of the Cenozoic age due to the siliceous pentasters (Fensome et al. 1993). Cell has a centrally located nucleus enclosed by a perinuclear strengthening layer and the curved arms of two pentasters. Cingulum is median and may be displaced one cingulum width. Sulcus is narrow in the hypocone and reaches the antapex, and it continues to the apex. Cells lack external plates but have internal pentasters, two large ones enclosing the nucleus and smaller ones in the cytosol. Electron microscopy on the type species has revealed extrusive organelles with complex structures named *docidosomes*, nuclear pores located in depressions in the nuclear envelope and the structure of the flagellar apparatus (Hansen 1993).

Actiniscus canadensis

Bursa ex Carty (herein).

SYNONYM: *Actiniscus canadensis*. Bursa 1969, 414, Figs. 17, 18, 20–22.

Freshwater, athecate, with small internal siliceous pentaster. **Fig. 16**

Figure 16. *Actiniscus canadensis.* (a) Ventral view with centrally located nucleus (redrawn from Bursa 1969, Fig. 17) (b) Pentasters (redrawn from Bursa 1969, Figs. 16, 20).

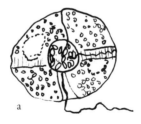
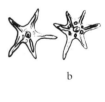

EXTERNAL FEATURES: Spherical cell; cingulum median without or with slight displacement; flagella present; sulcus in the hypocone with parallel sides reaching the antapex, continuing as a suture to the apex; reproduction by binary fission.

INTERNAL FEATURES: Nucleus centrally located; protoplast colorless; rudimentary, weakly silicified pentasters in some cells.

SIZE: 28–43 μm diameter.

ECOLOGICAL NOTES: Found in Great Bear Lake, NT; rare; seen only following sample sedimentation, never in net tows (Bursa 1969).

LOCATION REFERENCE: NT (Bursa 1969).

Amphidinium

Claparède and J. Lachmann 1859, 410. emend Jørgensen, Murray, and Daugbjerg 2004, 351–365.

TYPE SPECIES: *Amphidinium operculatum*. Claparède and J. Lachmann 1859, 410–411, Pl. 20, Figs. 9–10.

Cingulum divides cell into small finger-like epicone and larger hypocone.

Athecate, free swimming or benthic sand dwelling; brackish or marine; cingulum divides cell into finger-like epicone and large hypocone; dorsoventrally flattened. Species are determined by cell shape, size, proportions of epicone and hypocone, presence/absence of plastids and their color. Analysis of large-subunit ribosomal DNA, and the type species *A. operculatum*, redefined *Amphidinium* as a genus with a small, left-deflected epicone. Cladistic analysis grouped species with small finger-like epicones with the type species; species with larger epicones grouped with a *Gymnodinium* clade (Jørgensen et al. 2004). Some freshwater species were removed into the genus *Prosoaulax*, while others are considered not to be *Amphidinium* sensu stricto but will remain as such until further research places them in new genera (Calado and Moestrup 2005).

Key to species of *Amphidinium*

1a. Cell photosynthetic with blue-green plastids ....*A. wigrense*
1b. Cell photosynthetic with yellow-gold plastids, brackish ....*A. klebsii*
1c. Cell photosynthetic with yellow-gold plastids, freshwater ....*A. luteum*

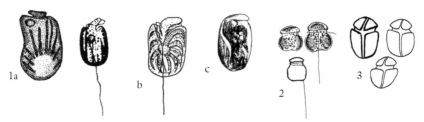

Plate 1. Species of *Amphidinium*. 1. *A. klebsii*. Note finger-like epicone. (a) Redrawn from Klebs 1884, Taf 10, Figs. 11–12 (b) Ventral view (Thompson 1950, Fig. 1) (c) Sketched from Barlow and Triemer 1988, Fig. 1. 2. *A. luteum*. Ventral and dorsal views. Note rounded epicone (Skuja 1939, Taf 10, Figs. 18–20, by permission). 3. *Amphidinium wigrense*. Three variations of ventral view (Wołoszyńska 1925a, Fig. 1).

Amphidinium klebsii

Kofoid and Swezy 1921, 144, Fig. U14.

BASIONYM: *Amphidinium operculatum* Claparède and Lachmann in Klebs 1884, Taf. 10, Figs. 11–12.

Amphidinium klebsii was considered a synonym of *A. operculatum* but differs from the type species by the presence of furrows on the cell surface (Murray et al. 2004).

Small epicone, brackish water, golden. **PL1(1)**

EXTERNAL FEATURES: Cell morphologically plastic, varying from rounded rectangular to round, slight dorsovental compression; cingulum divides cell into small finger-like epicone (<1/10 total cell length) and long hypocone; sulcus extends length of hypocone and is right of median. Sexual reproduction involved the fusion of two assimilative cells/gametes producing a planozygote that remained motile with two trailing flagella for a few weeks, nucleus became spherical, huge (>50% cell volume), exhibited nuclear cyclosis; eventually a cyst was produced (Barlow and Triemer 1988).

INTERNAL FEATURES: Protoplast with central pyrenoid, no eyespot; assimilative cells with crescent-shaped nucleus in posterior hypocone; chloroplast lobes peripheral and directed posteriorly (Barlow and Triemer 1988).

SIZE: 17–22 µm long by 10–14 µm wide, 7–10 µm thick (Thompson 1950).

ECOLOGICAL NOTES: Found in brackish water, saltwater marsh, ponds near the ocean.

LOCATION REFERENCE: MA (Herdman 1924), MD (Thompson 1950), NJ (Barlow and Triemer 1988).

Amphidinium luteum

Skuja 1939, 148–149, Taf. 10, Figs. 18–20.

Calado and Moestrup (2005) synonymized *A. luteum* with *Prosoaulax lacustris* but indicated that further research was needed.

Golden, small epicone, freshwater. **PL1(2)**

EXTERNAL FEATURES: Cell small, about 25% epicone and about 75% hypocone; lobed where sulcus reaches antapex.

INTERNAL FEATURES: Three to four parietal, discoid, yellow chloroplasts (Skuja 1939).

SIZE: 11–13 µm in length and width, 8–10 µm thick.

ECOLOGICAL NOTES: Reported as present all year from the experimental lakes area, northwestern Ontario (Kling and Holmgren 1972) and from the euplankton of a tundra pond, NT (Sheath and Hellebust 1978).

LOCATION REFERENCE: NT (Sheath and Steinman 1982, Sheath and Hellebust 1978), ON (Kling and Holmgren 1972), WA (Larson et al. 1998).

Amphidinium wigrense

Wołoszyńska 1925a, 3, 8, Fig. 1.

Blue-green plastids, freshwater. **PL1(3)**

EXTERNAL FEATURES: Inverted mushroom shape; wide, deeply incised cingulum divides cell into 25% epicone, 65% hypocone; sulcus to apex and antapex.

INTERNAL FEATURES: Two to seven blue-green plastids; no eyespot; nucleus in hypocone.

SIZE: 14–25 µm long by 10.5 µm wide at top of hypocone.

ECOLOGICAL NOTES: Collected (rare) from fall phytoplankton in a pond (Wilcox and Wedemayer 1985).

LOCATION REFERENCE: WI (Wilcox and Wedemayer 1985).

Bernardinium

Chodat 1923, 40–41.

TYPE SPECIES: *Bernardinium bernardinense* Chodat 1923, 41–42.

Colorless, with incomplete, right-descending cingulum.

Athecate, oval cell with firm envelope, free floating, motile, cingulum incomplete beginning in a midventral position then descending to the right, ending in a middorsal position; sulcus (difficult to see) narrow, extending from cingulum to antapex. Nucleus midcell; no eyespot (?); brown-red accumulation bodies (Chodat 1923, Javornický 1997).

Bernardinium bernardinense

Chodat 1923, 41–42, Fig. 7.

Nonphotosynthetic, athecate cell with incomplete, right-descending cingulum. **PL2(1)**

EXTERNAL FEATURES: Small oval cell with slight dorsoventral compression; distinctly incised cingulum begins in median position and descends to the right, curving around the cell and ending middorsally; lower epicone forms a protruding lip over the cingulum; sulcus extends partly into hypocone without reaching antapex.

INTERNAL FEATURES: Nucleus in hypocone; red eyespot in sulcal area; red-brown accumulation body in epicone.

SIZE: 11–14.5 μm long by 8–11 μm wide (Javornický 1997).

LOCATION REFERENCE: MN (Meyer and Brook 1969), TX (SC 830107-3).

COMMENTS: Reported by Thompson (1950) but diagrams reinterpreted as *Esoptrodinium gemma* by Javornický (1997). Without illustrations or descriptions of what researchers saw, it is difficult to assign reports of "*Bernardinium bernardinense*" to either *Bernardinium bernardinense* or *Esoptrodinium gemma,* so only Thompson's report will be moved, other reports of *Bernardinium bernardinense* will remain.

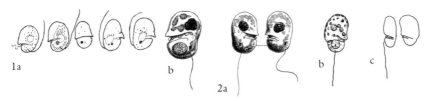

Plate 2. Athecate genera with incomplete cingulum. 1. *Bernardinium bernardinense.* Cingulum begins on the right side. (a) Series of drawings: origin of longitudinal flagellum on same side as origin of cingulum; presence of eyespot; last drawing shows end of cingulum on side opposite longitudinal flagellum (Chodat 1923, Fig. 7) (b) Nucleus in hypocone; eyespot in sulcus (Javornický 1997, Table 1, Fig. 1, by permission from E. Schweizerbart'sche Verlagsbuchhandlung). 2. *Esoptrodinium gemma.* Cingulum begins on the left side. (a) Dark accumulation body in epicone; eyespot in sulcus (Javornický 1997, Table 2, Figs. 1a, b, by permission from E. Schweizerbart'sche Verlagsbuchhandlung) (b) Ventral (c) Dorsal (Thompson 1950, Figs. 21–23).

Cystodinedria

Pascher 1944a, 380–381, Figs. 5–7.

TYPE SPECIES: *Cystodinedria inermis* (Geitler) Pascher 1944a, 380–381, Fig. 5.

Oval cell attached to filamentous green algae.

Cells sessile, attached along long axis, slight flattening on attached surface, poles broadly rounded, parasitic. Chromatophores may be green, orange, or brown depending on stage of metabolism. Work by Pfiester and Popovský (1979) and Popovský and Pfiester (1982) followed *Cystodinedria* as orange cells produced two amoebae with axopods that settled on *Spirogyra* and consumed several cells before transforming into the *Cystodinedria* shape. They are considered dinoflagellates because gymnodinioid zoospores have been reported. For additional discussion, see "Taxonomy" in the introduction.

Starmach (1974) lists 12 species; Popovský and Pfiester (1990) list one, *C. inermis*.

Cystodinedria inermis

(Geitler) Pascher 1944a, 380–381, Fig. 5.

BASIONYM: *Raciborskia inermis* Geitler 1943, 173, Abb 4

SYNONYM: *Cystodinium breviceps* Geitler 1928b, 81, Text Figs. 1, 2, 3, Taf. 7, Figs. 1, 2.

Oval, attached, parasitic on filamentous green algae. **Figs. 17, 18; CP2(1,2)**

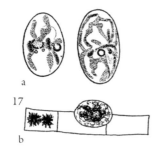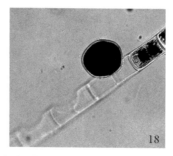

Figure 17. *Cystodinedria inermis.* (a) Individual cells (Pascher 1944b, Fig. 5e, f) (b) Cell in position on filament of *Zygnema*, empty cell contents below dinoflagellate cell (Carty 2003, Fig. 2H).

Figure 18. *Cystodinedria inermis* on *Zygnema* filament (stained with iodine; note the starch in the dinoflagellate and in the green alga stain the same), cell beneath *Cystodinedia* is empty.

EXTERNAL FEATURES: Oval to hemispherical cells, flattened on bottom.

INTERNAL FEATURES: Golden, brown, green, or orange cell contents.

SIZE: 18–48 μm long by 19–32 μm wide.

ECOLOGICAL NOTES: found on *Spirogyra, Zygnema, Oedogonium*.

Cystodinium

Klebs 1912, 441–442.

TYPE SPECIES: *Cystodinium bataviense* Klebs 1912, 377, Fig. 2A–K.

Nonmotile, planktonic, freshwater, golden, spindle- to lemon-shaped cells enclosed in a transparent, smooth covering; may have polar extensions.

Species have been distinguished by overall shape, size, and appearance of the poles. Starmach (1974) listed 25 species of *Cystodinium*. It has not always been clear what life stage was illustrated for a species or what feature was measured for length (assimilative cell, dividing cell, protoplast, or protoplast plus horns) so Thompson and Meyer (1984) recognized five species not including *C. cornifax,* which is synonymous with *C. iners.* Popovský and Pfiester (1990) reduced Thompson and Meyer's five species to three (*C. cornifax* for all taxa with terminal horns, *C. bataviense* for lunate cells with blunt horns, and *C. phaseolus* for elliptical to bean-shaped cells). See also discussion under "Taxonomy" in the introduction.

The life cycle includes an immobile assimilative stage where the protoplast fills the cell, and an immobile zoospore that is gymnodinioid in appearance (distinct cingulum and sulcus, red eyespot) and occurs when the protoplast contracts from the outer wall. Zoospores may divide into two; upon release these secrete a parental type envelope. The two division products may be flagellated and thecate, swim briefly, then lose the theca and transform into the parental type or, acting as gametes, fuse to form a round, thick-walled zygote (Pfiester and Lynch 1980 for *Cystodinium bataviense*). *Cystodinium bataviense* was also observed to divide into 10–15 μm light brown to colorless amoebae that parasitize filaments of *Oedogonium, Spirogyra,* and *Mougeotia,* eventually forming the parental shape (Pfiester and Lynch 1980).

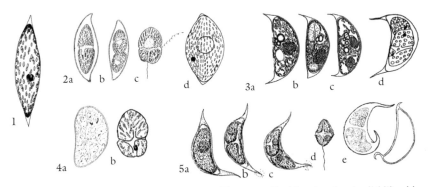

Plate 3. Species of *Cystodinium.* 1. *C. acerosum.* Elongate cell with pointed ends, disklike chloroplasts (Thompson and Meyer 1984, Fig. 3, with permission). 2. *C. bataviense.* Ends bluntly rounded. (a) Cell within envelope (b) Cell dividing (c) Motile cell (Klebs 1912, Figs. 2 A,F,J) (d) Cell with numerous disklike chloroplasts, eyespot, and central nucleus (Thompson 1949 Fig. 9). 3. *C. iners.* (a) Typical cell (b) Cell with one end less pointy (c) Cell contents dividing (Geitler 1928a Figs. 2a,b,c) (d) Cell with two division products (Thompson 1949, Fig. 7). 4. *C. phaseolus.* (a) Bean-shaped cell in envelope (b) Gymnodinioid cell with eyespot (Pascher 1928, Figs. 3a,d, 5a). 5. *C. steinii.* (a) Typical cell (b) Cell contains gymnodinioid within envelope (c) Two division products within envelope (d) Motile cell (Klebs 1912, Figs. 3A,B,C,F) (e) Cells showing dissimilarity of poles and cell torsion (Prescott 1951b, Pl. 93, Figs. 1, 2).

Key to species of *Cystodinium*

1a. Poles of cell rounded 2
1b. Poles of cell extended 3
2a. Cell bean-shaped .... *C. phaseolus*
2b. Cell lemon-shaped .... *C. bataviense*
3a. Cell fusiform, poles an extension of body .... *C. acerosum*
3b. Cell lunate, poles with hyaline spines 4
4a. Cell body 44–64 μm long .... *C. iners*
4b. Cell body 65–100 μm long .... *C. steinii*

Cystodinium acerosum

Thompson in Thompson and Meyer 1984, 88–89, Fig. 3.

Immobile, fusiform with gradually tapering poles. **PL3(1)**

EXTERNAL FEATURES: Cell within fusiform envelope, not extending into poles.

INTERNAL FEATURES: Brownish parietal plastids.

SIZE: Cell body 55–60 μm long by 15–18 μm thick, 67–72 μm long including processes.

ECOLOGICAL NOTES: Collected from a boggy pond, July and August (Thompson and Meyer 1984).

LOCATION REFERENCE: MD (Thompson and Meyer 1984).

Cystodinium bataviense

Klebs 1912, 377–381, Fig. 2A–K.

Large, immobile, planktonic, oval to lemon-shaped cell; poles bluntly rounded; may form surface film. **PL3(2); Figs. 19, 20; CP2(3–5)**

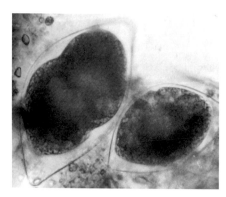

Figure 19. *Cystodinium bataviense.* Two cells, one showing division occurring, the other with distinctive cingulum.

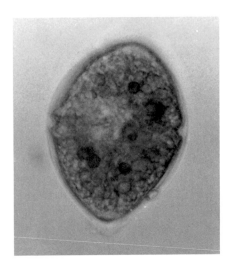

Figure 20. Single cell with accumulation bodies, distinctive cingulum.

EXTERNAL FEATURES: Lemon-shaped assimilative cell with both sides convex; asexual reproduction produces two *Gymnodinium*-like zoospores or immobile aplanospores, which when released form cells shaped like the parent; reproducing cells 87–130 μm long by 46–62 μm wide.

INTERNAL FEATURES: Numerous parietal golden-brown plastids; red eyespot; large red accumulation bodies sometimes.

SIZE: Protoplast 67–99 μm long by 45–49 μm wide.

ECOLOGICAL NOTES: Most reported collections were made during summer months (July–September) from small lakes; *Cystodinium* can produce surface film on ponds, and the entire surface will have a golden tinge caused by the positive phototaxis of the zoospore (Timpano and Pfiester 1985).

Cystodinium iners

Geitler 1928a, 5–6, Figs. 2, 3.

SYNONYMS: *Raciborskia inermis* Geitler 1943, 173–174, Fig. 4; *Cystodinium brevipes* pro parte Geitler 1928b, 68, Fig. 1; *Cystodinium cornifax* (Schilling) Klebs 1912, 442; *Cystodinium unicorne* Klebs 1912, 442; *Glenodinium cornifax* Schilling 1891a, 285, Pl. 10, Fig. 18.

Golden brown, immobile lunate cell with poles extended into slightly bent, asymmetric sharp points. **PL3(3); Fig. 21; CP2(6,7)**

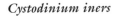

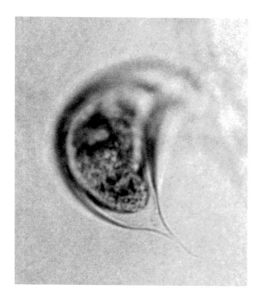

Figure 21. *Cystodinium iners* with terminal spines (one curved out of plane of focus).

EXTERNAL FEATURES: Immobile, floating, lunate cell; one side convex, the other flat to slightly concave or convex; pointed poles curve in different directions or different planes, giving an asymmetrical appearance. Similar to *C. steinii* but shorter.

INTERNAL FEATURES: Numerous golden-brown parietal plastids.

SIZE: Protoplast size 44–64 µm long by 22–30 µm wide, overall length (including horns) 61–81 µm.

ECOLOGICAL NOTES: Collected April, July to September, December from swampy, boggy ponds. I have collected it from squeezings of macrophytes, among desmids. Collected from epiphytic bromeliads in PR (Foerster 1971).

Cystodinium phaseolus

Pascher 1928, 241–254, Figs. 2–7.

Bean-shaped immobile cell. **PL3(4)**

EXTERNAL FEATURES: Bean-shaped immobile cell, convex on dorsal side, slightly indented on ventral, poles broadly rounded. Cell may produce two types of zoospores, a larger one that gives rise to a new individual or smaller ones that may function as gametes (Pascher 1928).

INTERNAL FEATURES: Plastids ribbon-like or discoidal; refractile bodies present; eyespot may be present.

SIZE: 13–53 µm long by 8–30 µm wide.

ECOLOGICAL NOTES: Collected from small mesotrophic lake (Meyer and Brook 1969).

LOCATION REFERENCE: MN (Meyer and Brook 1969).

Cystodinium steinii

Klebs 1912, 381, Fig. 3, Taf. 10, 2a, b.

Large, lunate, immobile cell with poles extended into spines that are not in the plane of the cell. **PL3(5)**

EXTERNAL FEATURES: Floating, immobile cell, one side convex, the other flat to concave, spines at poles twisted in different directions. Similar to *C. iners* but larger.

INTERNAL FEATURES: Eyespot may be present.

SIZE: 65–100 µm long, 25–35 µm diameter.

ECOLOGICAL NOTES: It was rare in lakes and acid swamps (Prescott 1951b).

Dinamoeba

Pascher 1916a, 118–128.

SYNONYM: *Dinamoebidium* Pasher 1916b, 31.

TYPE SPECIES: *Dinamoeba varians* Pascher 1916a, 118–128, Taf. 10, Figs. 1–28.

Assimilative cells amoeboid.

Freshwater or marine amoeboid cells, planktonic or attached, cingulum present, no sulcus, no eyespot, division of amoeboid cell into two similar cells or production of gymnodinioid zoospores (Popovský and Pfiester 1990). The name *Dinamoeba* has priority (Fensome et al. 1993, 165).

Dinamoeba coloradense

(Bursa) Carty comb. nov.

BASIONYM: *Dinamoebidium coloradense* Bursa 1970, 146, Figs. 1–8.

Freshwater, gymnodinioid cells attached to rocks in fast-moving streams. **Fig. 22**

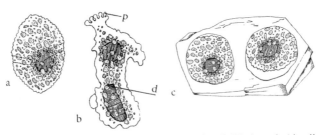

Figure 22. *Dinamoeba coloradense.* (a) Gymnodinioid cell (b) Amoeboid cell, where p = pseudopodia, d = ?diatom (c) Attached to sand grains (Bursa 1970, Figs. 1, 5, 8).

EXTERNAL FEATURES: Metabolically plastic gymnodinioid cells with median cingulum more or less indented; flattened ventral surface; sessile cells move in amoeboid manner, extending short, blunt pseudopodia. Cell division during amoeboid stage by lengthwise fission (Bursa 1970).

INTERNAL FEATURES: Protoplast with irregularly scalloped edges, numerous yellow-gold plastids; no eyespot; no flagella seen; no vacuoles (Bursa 1970).

SIZE: 19–38 μm long by 13–23 μm wide, amoeba 28–46 μm diameter.

ECOLOGICAL NOTES: Amoebae ingest algae by phagocytosis; collected (rare) from rock scrapings, sand grains, and *Fontinalis* (Bursa 1970).

LOCATION REFERENCE: CO (Bursa 1970).

Dinastridium

Pascher 1927, 27–32.

TYPE SPECIES: *Dinastridium sexangulare* Pascher 1927, 27, Figs. 25–29.

Nonmotile, irregularly polygonal cell with six angles bearing short spines.

Free floating, nonmotile, cells with six (5–7) angles to the wall, each angle with one or two short spines; plastids discoid, parietal; reproduction by gymnodinioid-like zoospores and autospores. The several species are considered the hypnospores of other taxa (Popovský and Pfiester 1990); see discussion under "Taxonomy" in the introduction.

Dinastridium sexangulare

Pascher 1927, 27, Figs. 25–29.

Irregularly hexagonal golden cell with horns at angles. **Figs. 23, 24**

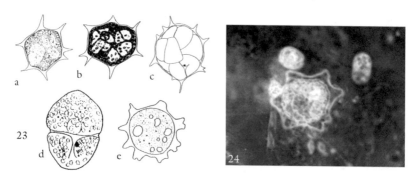

Figure 23. *Dinastridium sexangulare.* (a, b) Cells (c) Cell with swarmers (d) Individual swarmer cell (Pascher 1927, Fig. 25a,b, 27b, 27d) (e) Cell with more irregular processes (Forest 1954a, Fig. 508, used with permission).

Figure 24. *Dinastridium sexangulare* with irregular lobes (Carty 2003, Fig. 10I).

EXTERNAL FEATURES: Outer wall of cell with irregular short blunt spines; round protoplast; thought to be the hypnospore of *Gymnodinium chiatosporum* (Popovský and Pfiester 1990).

INTERNAL FEATURES: Centrally located nucleus.

SIZE: About 40 µm diameter.

ECOLOGICAL NOTES: Found in a pond and a swamp (TN, Forest 1954a).

LOCATION REFERENCE: TN (Forest 1954a), MD (SC 060623), TX (SC Brazos Co).

Dinococcus

Fott 1960, 149.

SYNONYM: *Raciborskia* Wołoszyńska 1919, 199.

TYPE SPECIES: *Dinococcus bicornis* (Wołoszyńska) Fott 1960, 149.

Attached cell with single spine at each end.

Epiphytic; freshwater cell attached with short stipe; plastids numerous, arranged around a pyrenoid. Reproduction by gymnodinioid zoospores. Although *Raciborskia* Wołoszyńska is an older name for this taxon, *Raciborskia* is also the genus of a myxomycete (Fensome et al. 1993). Species were transferred to *Dinococcus* by Fott and others; there are two species in Starmach (1974), three in Popovský and Pfiester (1990), though with a note indicating probably only one species with variations.

Dinococcus bicornis

(Wołoszyńska) Fott 1960, Preslia 32:150.

BASIONYM: *Raciborskia bicornis* Wołoszyńska 1919, 199, Taf. 14, Figs. 15–17.

Attached cell with single spine at each end. **Fig. 25; CP3(1)**

EXTERNAL FEATURES: Nonmotile, attached oval cell with single spine at each end.

SIZE: 25–35 µm long (includes spines), 9–12 µm diameter (Prescott 1951b).

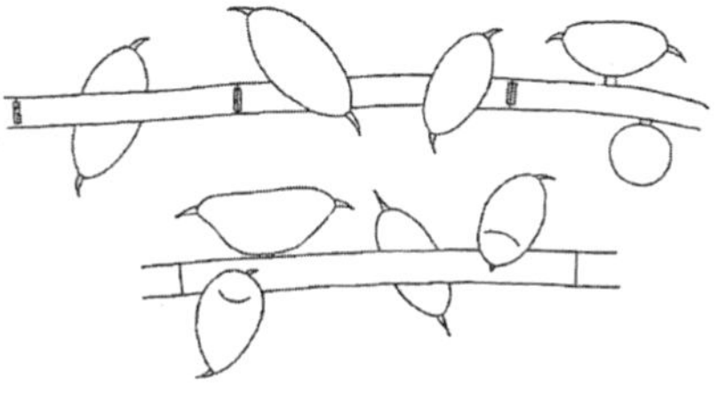

Figure 25. *Dinococcus bicornis.* Different views of cells attached to filamentous algae (Wołoszyńska 1919, Pl. 14, Fig. 15).

ECOLOGICAL NOTES: Cells attach to filamentous algae and develop from gymnodinioid swarmers lacking eyespots (Wołoszyńska 1919). Rare, attached to aquatic moss in 35 ft. water (Prescott 1951b), or to filamentous algae.

LOCATION REFERENCE: MN (Meyer and Brook 1969 as *Raciborskia bicornis*), TN (Johansen et al. 2007), WI (Prescott 1951b as *Raciborskia bicornis*), Panama (Prescott 1955).

Esoptrodinium

Javornický 1997, 35–36.

TYPE SPECIES: *Esoptrodinium gemma* Javornický 1997, 36–38, Table 2.
Heterotrophic cell with incomplete, left-descending cingulum.

Oval, athecate, slightly dorsoventrally compressed cells with incomplete median cingulum that begins at its junction with a weak sulcus and descends on the left side to a middorsal position. Nucleus in hypocone; red eyespot may be present in sulcus; may have chloroplasts; may have accumulation body in epicone; reproduction by binary fission. *Esoptrodinium* is a mirror image of *Bernardinium bernardinense* Chodat, which has a right-descending incomplete cingulum (Javornický 1997).

Esoptrodinium gemma

Javornický 1997, 36–38, Table 2.

BASIONYM: *Bernardinium bernardinense* in Javornický 1962, 98–113, Table 7.
Heterotrophic cell, incomplete, left-descending cingulum. **PL2(2); CP2(8,9)**

EXTERNAL FEATURES: Elliptical cell; epicone and hypocone broadly rounded; some dorsoventral compression; cingulum incomplete, dividing cells into larger epicone and smaller hypocone; margin of epicone bordering the cingulum is slightly extended as a lip. Reproduction by division during the motile stage (Javornický 1962).

INTERNAL FEATURES: The cell may contain colored (yellow, brown) accumulation bodies in the epicone; spherical nucleus in the hypocone; diffuse, parietal yellow-green plastids (Thompson 1950). Recent work found chloroplasts and an eyespot in addition to a peduncle and feeding veil (Calado et al. 2006). Cells from ponds in NC had large red accumulation bodies and were maintained in culture by feeding them *Cryptomonas* (Fawcett and Parrow 2012); cells collected in MD from a puddle were ingesting *Chlamydomonas* (Delwitch pers. comm.).

SIZE: 15–17 μm long by 10–12 μm wide.

LOCATION REFERENCE: KS (Thompson 1950), NC (Fawcett and Parrow 2012), MD (Delwitch, pers. comm.), TN (SC 040713-6).

Gloeodinium

Klebs 1912, 411–416, 445.

TYPE SPECIES: *Gloeodinium montanum* Klebs 1912, 445, Fig. 13.

Round, nonmotile, golden cells in (thick) (laminated) gelatinous sheath, associated with *Sphagnum* moss.

Round, photosynthetic cells in twos and fours in homogeneous or laminated envelope. *Gloeodinium* is considered by some to be a stage in the life cycle of *Hemidinium*, as *Gloeodinium* produces *Hemidinium*-like zoospores and *Hemidinium* produces a *Gloeodinium*-like resting stage. Molecular work using rDNA placed *Gloeodinium* and *Hemidinium* distant from each other (Logares et al. 2007a). Examination of location reports of either rarely includes the other, so I am retaining both genera.

Gloeodinium montanum

Klebs 1912, 445, Fig. 13, p. 414.

Round, nonmotile, golden cells in (thick) (laminated) gelatinous sheath, associated with *Sphagnum*. **Figs. 26, 27; CP3(2)**

EXTERNAL FEATURES: Round, immobile, golden cells; in gelatinous sheath that may be laminated, or may be in packets of cells; reproduction by *Hemidinium*-like zoospores. Reproduction occurs when assimilative cells divide to produce similar cells that remain in the colonial envelope; alternatively two or four

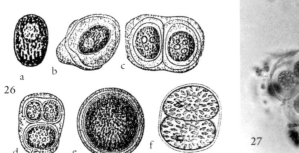

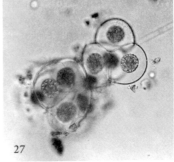

Figure 26. *Gloeodinium montanum.* (a) Single cell (b) Cell within asymmetric sheath (c) Cells within concentric sheaths (d) Cell following division (e) Rounder cell (Klebs 1912, Fig. 13 A, B, C, F, G) (f) Two cells in single sheath (Thompson 1949, Fig. 1).

Figure 27. *Gloeodinium montanum* with wide sheath around cells, cells in clump, some in pairs (Carty 2003, Fig. 10J).

swarmers are produced that may become motile or re-form assimilative cells (Popovský 1971a, Kelley and Pfiester 1989). Swarmers are capable of acting like zoospores or gametes; as gametes they fuse to form large aplanospores. Swarmers are short lived, with fewer plastids and starch grains compared with assimilative cells. Hyponozygotes may also be formed (Kelley and Pfiester 1989).

INTERNAL FEATURES: Golden chloroplasts.

SIZE: Cells 18–28 μm diameter, colony 69–74 μm diameter.

ECOLOGICAL NOTES: Reliably associated with *Sphagnum* (even baled), free in the plankton or associated with aquatic plants in sphagnum bogs.

Gymnodinium

Stein 1883, emend G. Hansen and Moestrup in Daugbjerg et al. 2000.

TYPE SPECIES: *Gymnodinium fuscum* (Ehrenberg) Stein 1878, 89.

Median cingulum, motile, athecate.

Genus is distinguished from similar athecate taxa by a median cingulum (*Amphidinium* has a supramedian cingulum, *Katodinium* has a submedian cingulum, *Gyrodinium* has a cingulum with ends greatly offset), cells lack a cellulose wall but do have a periplast. The emended description includes a horseshoe-shaped apical groove beginning near the top of the sulcus and almost encircling (counterclockwise) the apex, and ultrastructural features of the nucleus (Daugbjerg et al. 2000). Phagotrophy is known in this genus and may be indicated in cells lacking dinoflagellate pigmentation. Species are separated based on overall shape; size; shape and position of nucleus; plastid presence, shape, and arrangement; presence of an eyespot; accumulation bodies; and features of the sulcus. An important feature is the junction of the cingulum and sulcus. In some species, when the sulcus enters the epicone, the lower margin of the upper right quadrant may project into the sulcus. Species determination can be difficult (except for the distinctive species such as *G. aeruginosum, G. fuscum*) because cells round up under stress and overall shape and sulcus features are lost. There are probably more than 100 species of freshwater *Gymnodinium* in the world; 51 species are listed in Starmach (1974). Popovský and Pfiester (1990) synonymized many species of *Gymnodinium*, but inspection of line drawings shows that most of the synonymized species do not match the original description (especially in details of the sulcus). The revision of the genus keeps species with known apical grooves, but those lacking grooves and those for which an apical groove is unknown are currently unplaced. If a species has been identified as *Gymnodinium,* that designation will continue here unless a specific species is known to be transferred to a different genus. If the species you see does not match the species in this key, try looking first at some similar genera (*Amphidinium, Prosoaulax, Katodinium,*

Opisthoaulax), then at some of the thin-walled thecate species (*Glenodiniopsis, Jadwigia, Tovellia, Woloszynskia*). If you still don't find it, check other keys (Starmach 1974), and then think about naming a new species.

Key to species of *Gymnodinium*

1a. Cell colorless or color other than yellow gold 2
1b. Cell with yellow-gold plastids 9
2a. Plastids blue green, blue, or green 3
2b. Plastids lacking, eyespot lacking 7
3a. Cells green 4
3b. Cells blue or blue green 5
4a. Cells ovate, about 30 μm long, with eyespot .... *G. viride*
4b. Cells may have one end tapered, about 17 μm long .... *G. varians*
5a. Hypocone tapered to point, epicone rounded to conical .... *G. acidotum*
5b. Hypocone rounded to flattened 6
6a. Sulcus enters epicone, almost reaches antapex .... *G. aeruginosum*
6b. Sulcus only slightly into hypocone .... *G. limneticum*
7a. Colorless cell, weak cingulum and sulcus, may have colored bodies ....
.... *G. albulum*
7b. Colorless cell, distinct cingulum and sulcus 8
8a. Cruciate cingulum and sulcus, rounded hypocone .... *G. thompsonii*
8b. Cingulum deep, tapered hypocone, pink cytoplasm .... *G. helveticum*
9a. Posterior three lobed, small cell .... *G. triceratium*
9b. Posterior otherwise 10
10a. Posterior tapered to a tip, large cell 11
10b. Posterior rounded .... 12.
11a. Posterior tapered to curved tip, >100 μm long, *Sphagnum* bogs ....
.... *G. caudatum*
11b. Posterior broadly rounded, 50–80 μm long .... *G. fuscum*
12a. Found under ice .... *G. cryophilum*
12b. Found in open water 13
13a. Cell small, <20 μm long, no eyespot 14
13b. Cell larger 15
14a. Cingulum submedian, nucleus in epicone .... *G. luteofaba*
14b. Cingulum median .... *G. lacustre*
15a. Cell large, 80–90 μm long .... *G. mirabile*
15b. Cell smaller 16
16a. Cell 40–60 μm long 17
16b. Cell 20–40 μm long 18
17a. Cingulum submedian, cells slim, 27–37 μm wide .... *G. palustre*
17b. Cingulum median, cell broader 38–42 μm wide, posterior lobed ....
.... *G. uberrimum*
18a. Hypocone bilobed, alkaline habitats .... *G. excavatum*
18b. Hypocone rounded in ventral view 19
19a. Hypocone larger than epicone .... *G. inversum*
19b. Hypocone and epicone about the same size 20

20a. Left side of hypocone wavy .... *G. undulatum*
20b. Left side of hypocone smooth 21
21a. Sulcus only slightly into hypocone.. *G. skvortzowii*
21b. Sulcus at least halfway down hypocone 22
22a. Numerous brownish plastids .... *G. bogoriense*
22b. Four to six golden chromatophores.................................... *G. marylandicum*

Gymnodinium acidotum

Nygaard 1949, 155–156, Fig. 95.

Blue green, hypocone extended into short tail.
PL4(1); CP4(1)

EXTERNAL FEATURES: Epicone conical to hemispherical; hypocone extended into short tail; cingulum slightly supramedian; sulcus extends to antapex
and enters epicone; there is a short projection of the
epicone into the sulcus; cell dorsoventally flattened.

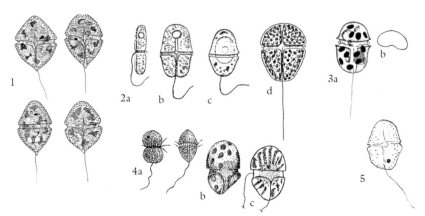

Plate 4. Blue, green, and blue-green species of *Gymnodinium*. 1. *G. acidotum*. Ventral and
dorsal views; note pointed antapex, slight penetration of sulcus into epicone (Nygaard 1949,
Fig. 95, with permission). 2. *G. aeruginosum*. (a) Side view showing dorsoventral compression
(b) Ventral view with rounded to almost bilobed posterior (c) Dorsal view with rounded posterior (Stein 1883, Taf 2, Figs. 19–21; note transverse flagellum depicted as cilia) (d) Ventral view
with numerous chloroplasts, deep sulcal penetration into epicone and rounded posterior
(Thompson 1950, Fig. 12). 3. *G. limneticum*. (a) Ventral view; note large chloroplasts, no sulcal penetration into epicone (b) Cross section showing dorsoventral compression (Lackey
1936, Pl 1, Figs. 8, 9, with permission). 4. *G. varians*. (a) Ventral (Maskell 1887; note transverse flagellum depicted as cilia) (b) Ventral (Popovský 1971b) (c) Ventral (Klebs 1912 as
Gymnodinium minimum). 5. *G. viride*. Ventral view, with eyespot (Penard 1891, Pl. 4, Fig. 11).

INTERNAL FEATURES: Nucleus in epicone; no eyespot; may contain red accumulation bodies (Nygaard 1949).

SIZE: 30–40 µm long by 20–30 µm wide.

ECOLOGICAL NOTES: Collected from a bloom in a bayou (Farmer and Roberts 1990), a bloom in a lake (Fields and Rhodes 1991).

COMMENTS: Cell color due to ingestion of *Chroomonas*. Pigmented cells in unialgal culture able to divide, but eventually nonpigmented cells result from one daughter cell not receiving a plastid or by digestion of plastid (and formation of accumulation body). Staining shows a dinoflagellate nucleus in the epicone of colorless cells and a second nucleus in the hypocone of pigmented cells (Fields and Rhodes 1991). Molecular analysis shows *G. acidotum* to belong to the emended *Gymnodinium* (Daugbjerg et al. 2000).

Gymnodinium aeruginosum

Stein 1883, Taf. 2, Figs. 19–21.

Blue-green, ovate cell; many plastids; flattened to bilobed posterior. **PL4(2); CP4(2,9)**

EXTERNAL FEATURES: Ovate motile athecate cells with hypocone tapered to slightly flattened antapically (may appear bilobed); epicone broadly rounded; strong dorsoventral compression; indented cingulum slightly above median, without displacement; sulcus extends into epicone and hypocone; epicone does not form a projection into the sulcus.

INTERNAL FEATURES: Small, numerous, peripheral, radially arranged blue-green chromatophores; accumulation bodies may be present; no eyespot.

SIZE: 20–36 µm long by 13–25 µm wide.

ECOLOGICAL NOTES: Cells swim in a straight path, quickly, until they make contact with something, there may be a few exploratory bumps, then cells continue swimming; widespread.

COMMENTS: Cells identified as *G. acidotum*, but looking like *G. aeruginosum*, were determined to be achlorotic but harboring a blue-green cryptomonad (Wilcox and Wedemayer 1984). Popovský and Pfiester (1990) synonymized *G. acidotum* with *G. aeruginosum*, but the overall shapes of the two taxa are different.

Gymnodinium albulum

Lindemann 1928, 292, Figs. 8–10.

Colorless, small. **PL5(1)**

EXTERNAL FEATURES: Small, athecate oval cell; epi-cone wider than hypocone; posterior flattened; cingu-lum median, weakly indented; sulcus extends into hypocone.

INTERNAL FEATURES: Without plastids; with red ac-cumulation body and refractive granules in cytoplasm; no eyespot.

SIZE: 12–18 µm long by 9–13 µm wide.

ECOLOGICAL NOTES: Collected from a swamp, February and March (Thompson 1947). Large number of reports from EPA documents.

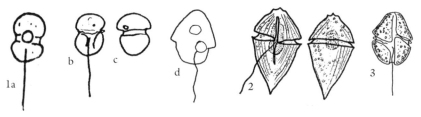

Plate 5. Nonphotosynthetic species of *Gymnodinium*. 1. *G. albulum*. (a) Ventral view (Linde-mann 1928, Fig. 8) (b) Ventral view (c) Dorsal view (redrawn from Thompson 1947, Pl 1, Figs. 1, 2) (d) Ventral (orig.). 2. *G. helveticum;* note scalloped apex, pointed antapex (a) Ven-tral (b) Dorsal (Penard 1891, Pl. 5, Figs. 10, 11) 3. *G. thompsonii;* note cruciate arrangement of cingulum and sulcus (Thompson 1950, Fig. 104).

Gymnodinium bogoriense

Klebs 1912, 439, Fig. 7 C, D.

Oval cell, brownish chloroplasts. **PL7(8)**

EXTERNAL FEATURES: Cell oval; dorsoventrally compressed; sulcus does not enter epicone and slants toward antapex with parallel sides; the right ventral postcingular portion greater than the left.

INTERNAL FEATURES: Dark brown chloroplasts arranged radially; no eyespot; nucleus central.

SIZE: 20 µm long by 16 µm wide (Klebs 1912).

LOCATION REFERENCE: Costa Rica (Umaña Villalobos 1988).

Gymnodinium caudatum

Prescott 1944, 371, Pl. 4, Figs. 14–16.

Large, golden, with eyespot and tail-like extension of hypocone. **PL6(2); CP4(3)**

EXTERNAL FEATURES: Ice cream cone–shaped; epicone broadly rounded; hypocone tapered to a curved tail; dorsoventrally compressed; incised cingulum medial and displaced; sulcus slightly into the epicone and halfway down the hypocone.

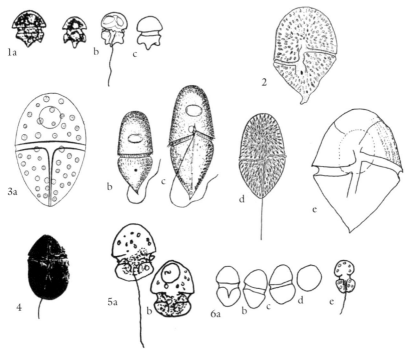

Plate 6. Yellow-golden species of *Gymnodinium*. 1. *G. triceratium;* note trilobed posterior (a) Two cells showing variability in lobing (Skuja 1939, Taf 10, Figs. 35, 37, with permission) (b) Ventral (c) Dorsal (redrawn from Thompson 1947, Pl. 1, Figs. 5, 6). 2. *G. caudatum* ventral view; note antapex extended into short tail, eyespot in sulcus (Prescott 1951b, Pl. 90, Fig. 1). 3. *G. fuscum.* (a) note rounded apex and antapex (Ehrenberg 1838, Pl. 22, Fig. 15) (b) Dorsal (c) Ventral; note posterior coming to a point (Stein 1883, Taf 2, Figs. 14–15; note transverse flagellum depicted as cilia) (d) Ventral view with numerous, radially arranged chloroplasts (Thompson 1947, Pl 1, Fig. 17) (e) Ventral view based on understanding of circum-apical groove, location of nucleus in epicone (based on images in Hansen et al. 2000). 4. *G. cryophilum* ventral view showing sulcus extending into epicone and hypocone (Wedemayer et al. 1982, Fig. 1; reprinted with permission from the Phycological Society of America). 5. *G. luteofaba* ventral views showing nucleus (nu) in epicone, single large chloroplast in hypocone (sketch based on Javornický 1965, Figs. 8, 10). 6. *G. lacustre.* (a) Ventral (b, c) Dorsal (d) Cross section showing no dorsoventral compression (Schiller 1933, Fig. 383) (e) Ventral view (orig.).

INTERNAL FEATURES: Numerous golden plastids; eyespot in sulcus in hypocone.

SIZE: 65–70 µm diameter, 104–118 µm long (Prescott 1944).

ECOLOGICAL NOTES: Collected from *Sphagnum* bog.

LOCATION REFERENCE: AK (SC 060713-2), WI (Prescott 1951b)

Gymnodinium cryophilum

(Wedemayer, Wilcox, and Graham) G. Hansen and Moestrup in Daugbjerg et al. 2000, 305, 317.

BASIONYM: *Amphidinium cryophilum* Wedemayer, Wilcox, and Graham 1982, 13–14, Fig. 1.

Cold loving, golden yellow. **PL6(4)**

EXTERNAL FEATURES: Morphology plastic but generally oval to elliptical with indented cingulum dividing cell into about one-third epicone, two-thirds hypocone; slight displacement of cingulum; sulcus narrow, extending as slit about halfway down hypocone and almost to the apex in the epicone.

INTERNAL FEATURES: With peduncle; numerous peripheral golden-yellow plastids; no eyespot; central nucleus; cell division occurs while cells motile. (Wedemayer et al. 1982).

SIZE: Average 33 µm long by 22 µm wide.

ECOLOGICAL NOTES: Collected in winter, sometimes dominant phytoplankter (3500 cells/mL), present under ice.

LOCATION REFERENCE: WI (Wedemayer et al. 1982)

Gymnodinium excavatum

Nygaard 1945, 52, Fig. 20.

Also in Nygaard 1949, 156, Fig. 96.

Bilobed posterior, finger-like projection of epicone into sulcus. **PL7(4)**

EXTERNAL FEATURES: Rounded epicone; hypocone bilobed; wide cingulum; sulcus to antapex and not entering epicone; finger-like extension of epicone into sulcus to level of hypocone.

INTERNAL FEATURES: Nucleus central; may have eyespot; oval, numerous, brownish yellow plastids.

SIZE: Cells 26–42 µm long by 21–34 µm wide.

ECOLOGICAL NOTES: Found in alkaline habitats (pH 7.3–8.6) often with *Trachelomonas volvocina* (Nygaard 1945, 1949).

LOCATION REFERENCE: Labrador (Duthie et al. 1976).

Gymnodinium fuscum

(Ehrenberg) Stein 1878, 95.

Gymnodinium fuscum (Ehrenberg); Stein 1883, Figs. 11–18 (illustration).

BASIONYM: *Peridinium fuscum* Ehrenberg 1834, 270.

Large, golden brown, ice cream cone–shaped cell; no eyespot. **PL6(3); CP4(4)**

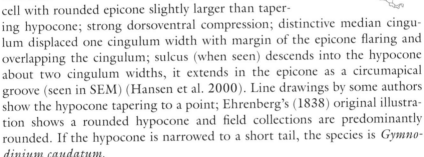

EXTERNAL FEATURES: Athecate, free-swimming cell with rounded epicone slightly larger than tapering hypocone; strong dorsoventral compression; distinctive median cingulum displaced one cingulum width with margin of the epicone flaring and overlapping the cingulum; sulcus (when seen) descends into the hypocone about two cingulum widths, it extends in the epicone as a circumapical groove (seen in SEM) (Hansen et al. 2000). Line drawings by some authors show the hypocone tapering to a point; Ehrenberg's (1838) original illustration shows a rounded hypocone and field collections are predominantly rounded. If the hypocone is narrowed to a short tail, the species is *Gymnodinium caudatum*.

INTERNAL FEATURES: No eyespot; no accumulation bodies; many golden-brown radially arranged plastids; nucleus in epicone.

SIZE: 50–80(100) µm long by 50–60 µm wide.

ECOLOGICAL NOTES: Found in the littoral of lakes and ponds in cool seasons (Whitford and Schumacher 1969), lily ponds and acid bogs (Prescott 1951b), acid marsh (Dawes and Jewett-Smith 1985).

Gymnodinium helveticum

Penard 1891, 58–59, Pl. 5, Figs. 10–16.

Apex scalloped; nonphotosynthetic; with colored contents (may be pink). **PL5(2)**

EXTERNAL FEATURES: Ice cream cone–shaped cell; epicone smaller than hypocone; slight dorsoventral compression; tapering to truncate (or slightly scalloped) apex in epicone, tapering to a point in the hypocone; deeply incised cingulum above median, without displacement; lower rim of epicone overhanging cingulum; sulcus (if visible) vertical, extending into epicone and hypocone without reaching apex or antapex.

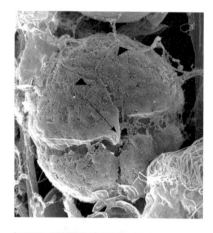

Figure 28. *Gymnodinium inversum.* Ventral view, circum-apical groove evident (arrowhead); note extension of left epitheca into sulcal area (arrow).

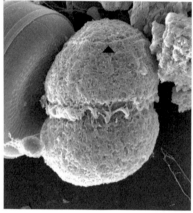

Figure 29. Side view with circum-apical groove (arrowhead) and transverse flagellum.

INTERNAL FEATURES: Nucleus in hypocone; no eyespot; without plastids but with colored contents, may have red accumulation bodies in epicone, green in hypocone; known to ingest green algae, flagellates, diatoms, and cyanobacteria (Irish 1979, Popovský 1982) which distort the shape.

SIZE: Cell 31 µm long by 21 µm wide.

ECOLOGICAL NOTES: Dominant in an ultraoligotrophic Arctic lake (Kalff et al. 1975). A *Gymnodinium*, thought to be *G. helveticum*, was found ingesting species of *Cyclotella* and *Stephanodiscus;* it was collected in less disturbed areas of Lake Michigan and Lake Superior at all depths and seasons (Frey and Stoermer 1980).

Gymnodinium inversum

Nygaard in Berg and Nygaard 1929, 294–295 Pl. 5, Figs. 28–36.

Hypocone larger. **PL7(5); Figs. 28, 29; CP4(5,6)**

EXTERNAL FEATURES: Oval to ovate cell; slightly dorsoventrally compressed; cingulum above median dividing cell into larger hypocone and smaller epicone, both rounded; cingulum slightly offset; sulcus acutely into epicone, narrow in hypocone descending vertically to antapex; right, ventral epicone forms an acute lobe into the sulcus. Based on SEM micrographs this is a true *Gymnodinium* with an apical groove.

INTERNAL FEATURES: Numerous round to oval brown plastids; no eyespot; nucleus in epicone.

SIZE: 27–36 µm long by 22–29 µm wide (Nygaard 1949).

ECOLOGICAL NOTES: Common in *Utricularia* mats (Hilgert 1976), Berg and Nygaard (1929) found a large population in waters 5.5 °C.

Gymnodinium lacustre

Schiller 1933, 374–375, Fig. 383.

SYNONYM: *Gymnodinium profundum* Schiller 1933 ad interim 399, Fig. 416.

Small, yellow. **PL6(6)**

EXTERNAL FEATURES: Cell oval; cingulum wide but may be difficult to see; sulcus narrow, not reaching antapex, indistinct.

INTERNAL FEATURES: No eyespot; accumulation bodies (red, colorless) may be present; plastids may be present.

SIZE: 9–25 µm long by 8–18 µm wide.

ECOLOGICAL NOTES: Found dominant in a sewage-polluted Arctic lake (Kalff et al. 1975).

Gymnodinium limneticum

Lackey 1936, 497, Pl. 1, Figs. 8, 9.

Oval, several blue plastids. **PL4(3)**

EXTERNAL FEATURES: Oval-elliptical cells; broadly rounded epicone and hypocone; wide cingulum slightly above median (hypocone larger than epicone), not displaced; sulcus not in epicone, extending halfway down hypocone; dorsoventral compression; indented ventral face.

INTERNAL FEATURES: Nucleus in hypocone; 8–12 oval blue plastids; no mention of an eyespot and none in the illustration (Lackey 1936).

SIZE: 25–35 μm long by 18–20 μm wide (Forest 1954a).

ECOLOGICAL NOTES: Most common dinoflagellate, September (Lackey 1936); rare in ponds, summer (NC, Whitford and Schumacher 1969).

Gymnodinium luteofaba

Javornický 1965, 56–58, Figs. 8–11.

Small cell; single, yellow, bean-shaped plastid in hypocone. **PL6(5)**

EXTERNAL FEATURES: Cell ovoid without compression; poles rounded; wide cingulum divides cell into larger epicone and smaller hypocone; sulcus does not enter epicone and weakly furrows the hypocone.

INTERNAL FEATURES: A single, large, yellow, bean-shaped plastid is found in the hypocone; nucleus in epicone; cytosol colorless; no eyespot.

SIZE: Cells 10–11 μm long by 5–8 μm wide.

LOCATION REFERENCE: QC (Janus and Duthie 1979).

Gymnodinium marylandicum

Thompson 1947, 7–8, Pl. 1, Parts 7–9.

Small oval cell, plastids present. **PL7(9)**

EXTERNAL FEATURES: Oval cell; epicone slightly wider than hypocone; both apex and antapex flattened; dorsoventrally compressed; cingulum median, displaced about one cingulum width; smooth-walled oval cyst 18–24 μm long and 14–21 μm wide. A cell illustrated in side view shows a vertical sulcal profile and an acutely beveled hypocone. Another cell shows a double outer wall and two cells within (Thompson 1947).

INTERNAL FEATURES: No eyespot; 4–6 parietal, golden plastids; orange-red oil globules present.

SIZE: 17–22 μm long by 13–19 μm wide.

ECOLOGICAL NOTES: Collected February and March from under the ice (Thompson 1947).

LOCATION REFERENCE: MD (Thompson 1947).

Gymnodinium mirabile

Penard 1891, 56–57, Pl. 5, Figs. 1–7.

Large, photosynthetic cell. **PL7(1)**

EXTERNAL FEATURES: Large oval cell; dorsoventrally compressed; cingulum divides cell into a tapered, broadly truncate epicone and a hypocone that is flattened to slightly bilobed.

INTERNAL FEATURES: Plastids are green to yellow or brownish, radially arranged; eyespot rare; clear granulations in the cytoplasm; trichocysts visible (Penard 1891).

SIZE: 80–90 µm long by 65–75 µm wide, 40–50 µm thick.

LOCATION REFERENCE: NU (Kalff et al. 1975), NT (Sheath and Steinman 1982), ON (Kling and Holmgren 1972).

ECOLOGICAL NOTES: Found dominant in a sewage-polluted Arctic lake (Kalff et al. 1975).

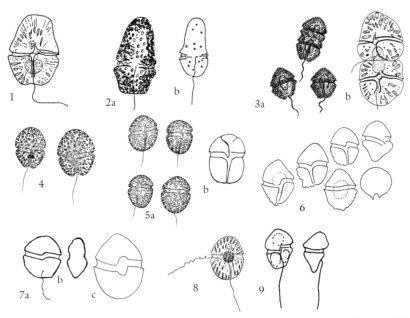

Plate 7. Yellow-golden species of *Gymnodinium*. 1. *G. mirabile* ventral view (Penard 1891, Pl 5, Fig. 1). 2. *G. palustre*. (a) Ventral view (Schilling 1891a) (b) Ventral view (Forest 1954a, Fig. 512). 3. *G. uberrimum*. (a) Ventral and dorsal views (Allman 1855, Pl 3, Figs. 9, 10, 14, reoriented) (b) Ventral (Penard 1891, Pl. 5, Fig. 8). 4. *G. excavatum* ventral views; note long extension into sulcus from lower epicone, no penetration of epicone by sulcus, central nucleus, eyespot, bilobed posterior (Nygaard 1949, Fig. 96, with permission). 5. *G. inversum*. Ventral and dorsal views. (a) Note slight penetration of epicone by sulcus, short extension of epicone into sulcus, rounded posterior, hypocone slightly larger than epicone (Nygaard 1949, Fig. 98, with permission) (b) Shape reconstructed from micrographs (orig.). 6. *G. undulatum* ventral, dorsal, and cross sections showing wavy margin of left hypocone (Wołoszyńska 1925b, Figs. 3E,F,G,I,J,K) 7. *G. skvortzowii*. (a) Ventral; note sulcus not reaching antapex (b) Side view (Skvortzov 1927, Fig. 1) (c) Ventral view (Hoham 1966, Pl. 19, Fig. 3, with permission) 8. *G. bogoriense* ventral view with sulcus extending to antapex (Klebs 1912, Fig. c). 9. *G. marylandicum* ventral and dorsal views (Thompson 1947, Pl. 1, Figs. 7, 8).

Gymnodinium palustre

Schilling 1891a, 277, Pl. 10, Fig. 11.

Elongate epicone; smaller rounded to flattened hypocone. **PL7(2)**

EXTERNAL FEATURES: Ovate cells; epicone tapered and rounded; hypocone broadly rounded; cingulum approximately median, without displacement; sulcus with narrow extension into epicone, wide in hypocone.

INTERNAL FEATURES: Round golden-brown plastids (Prescott 1951b). Molecular analysis shows *G. palustre* belongs to the emended *Gymnodinium* (Daugbjerg et al. 2000).

SIZE: 27–37 µm diameter, 40–60 µm long.

ECOLOGICAL NOTES: Collected from shallow marsh water, May and June (Prescott 1927), river (Forest 1954a).

Gymnodinium skvortzowii

Schiller 1933, 415, Fig. 435.

BASIONYM: *Gymnodinium hiemale* Skvortzov 1927, 123, Fig. 1.

Ovoid cell; strongly dorsoventrally compressed; sulcus weak. **PL7(7)**

EXTERNAL FEATURES: Cell ovoid; strongly dorsoventrally compressed; posterior flattened; cingulum median; sulcus not entering epicone, extending halfway into hypocone (Skvortzov 1927).

INTERNAL FEATURES: No eyespot; yellow chloroplasts (Skvortzov 1927).

SIZE: 28–31 µm long by 21–25 µm wide.

ECOLOGICAL NOTES: Rare in the winter plankton (Skvortzov 1927), possibly associated with *Dinobryon* (Hoham 1966).

LOCATION REFERENCE: MT (Hoham 1966).

Gymnodinium thompsonii

Kiselev 1954, 14.

SYNONYM: *Gymnodinium cruciatum* Thompson 1950, Figs. 103–104.

Cross-shaped intersection of cingulum and sulcus, nonphotosynthetic. **PL5(3); CP4(7,8)**

EXTERNAL FEATURES: Sulcus extends from apex to antapex; cingulum median, descending on both the left and right ventral face giving the appearance of displacement.

INTERNAL FEATURES: Nonphotosynthetic ovoid cell without eyespot; protoplast granular, gray, with refractive granules near periphery and food vacuoles (Thompson 1950).

SIZE: 19–21 µm long by 14–15.5 µm wide.

ECOLOGICAL NOTES: Collected from lake in July (Thompson 1950); cells with a distinctive cruciate cingulum and sulcus, 44 µm long by 33 µm wide, with a grainy cytoplasm, central nucleus, no eyespot, but golden (from recent meal?), collected from a shallow lake in March, Ohio.

LOCATION REFERENCE: KS (Thompson 1950), VA (Marshall and Burchardt 2004), OH (SC 040318-2).

Gymnodinium triceratium

Skuja 1939, 153, Table 10, Figs. 35–38.

Small golden cell, trilobed hypocone. **PL6(1)**

EXTERNAL FEATURES: Epicone rounded; hypocone two or three lobed; wide, slightly indented median cingulum, without displacement; sulcus (if seen) in hypocone only.

INTERNAL FEATURES: No eyespot; 1–5 golden plastids (mostly in epicone); may have red oil globules.

SIZE: 10–16 µm long by 9–13 µm wide, 7–9 µm thick.

ECOLOGICAL NOTES: Collected (common) in swamp pools (Whitford and Schumacher 1984).

Gymnodinium uberrimum

(Allman) Kofoid and Swezy 1921, 264–265.

BASIONYM: *Peridinium uberrima* Allman 1855, Pl. 3, Figs. 9–17.

Golden cell with eyespot. **PL7(3)**

EXTERNAL FEATURES: Rounded epicone; rounded to flattened hypocone; cingulum median without or with slight displacement; sulcus may enter epicone and penetrates to antapex.

INTERNAL FEATURES: Nucleus is central; golden plastids radiate; eyespot present; reddish brown granular contents, clear oil droplets (Allman 1855).

SIZE: 40–51 µm long and 38–42 µm wide (Kofoid and Swezy 1921).

ECOLOGICAL NOTES: Motile cells may form chains (Popovský and Pfiester 1990). Capable of forming blooms that discolor the water (Allman 1855). In samples collected from Lake Huron, individuals in the population were infected with virus-like particles and lacked a nucleus (Sicko-Goad and Walker 1979).

Gymnodinium undulatum

Wołoszyńska 1925b, 54 (Polish), 62 (German), Fig. 3E–K.

Left side of hypocone undulate. **PL7(6)**

EXTERNAL FEATURES: Hypocone slightly larger than epicone, left side wavy; cingulum slightly offset; sulcus not entering epicone, descending in hypocone with parallel side to antapex.

INTERNAL FEATURES: Chloroplasts golden brown, nucleus central.

SIZE: 35 µm long by 22 µm wide.

LOCATION REFERENCE: MT (Hoham 1966).

Gymnodinium varians

Maskell 1887, 7–8, Pl. 1, Fig. 9a, b.

Small green cell. **PL4(4)**

EXTERNAL FEATURES: Cell tiny; sulcus in hypocone only, reaches antapex; cell halves symmetrical, either tapered like a football or rounded (Maskell 1887).

INTERNAL FEATURES: Green, rod-shaped plastids in periphery of cell (Kofoid and Swezy 1921); an eyespot may be illustrated (Maskell 1887).

SIZE: 17 µm long.

LOCATION REFERENCE: SK (Hammer et al. 1983).

Gymnodinium viride

Penard 1891, 55–56, Pl. 4, Figs. 11–24.

Green, ovate, with eyespot. **PL4(5)**

EXTERNAL FEATURES: Oval cell; apex and antapex flattened; median, deeply incised cingulum without displacement; sulcus barely enters epicone and hypocone.

INTERNAL FEATURES: Elongate red eyespot in upper sulcus; numerous blue-green, radially arranged, cone-shaped plastids; refractive granules in cytoplasm.

SIZE: 29–32 μm long by 22–25 μm wide (Thompson 1947).

ECOLOGICAL NOTES: Collected in July (Thompson 1947).

LOCATION REFERENCE: BC (Wailes 1934, Stein 1975), MD (Thompson 1947).

The following species of *Gymnodinium* have also been reported:

Gymnodinium agile

Kofoid and Swezy 1921.

VA (Bovee 1960), marine.

Gymnodinium incurvum Nygaard var. *elongatum* Nygaard.

MA (from under the ice: Wright 1964) (This is probably *G. inversum* Nygaard since there is no species *G. incurvum* Nygaard).

Gymnodinium rotundatum

Klebs 1912

Collected from under the ice in a pond (MA, Wright 1964), QC (Brunel and Poulin 1992). This species is in the marine genus *Heterocapsa*.

Gyrodinium

Kofoid and Swezy 1921, 273–278, emend G. Hansen and Moestrup in Daugbjerg et al. 2000, 312.

TYPE SPECIES: *Gyrodinium spirale* (Bergh) Kofoid and Swezy 1921.

Motile, athecate with cingulum angled downward.

Marine or freshwater planktonic, athecate (though with longitudinal striations), motile unicell. Cingulum begins at sulcus in upper right portion of cell, crosses left, descends across dorsal surface, and indents right side of cell at least one cingulum width below origin; sulcus usually depicted at an angle to cingulum. Detailed examination of the type species (marine) showed an elliptical bisected apical groove, complex nuclear envelope, distinct vertical ridges, and evidence of feeding on diatoms, dinoflagellates, and cryptomonads (Hansen and Daugbjerg 2004).

Gyrodinium pusillum

(Schilling) Kofoid and Swezy 1921, 329–330, Fig. CC, p. 3.

BASIONYM: *Gymnodinium pusillum* Schilling 1891a, 279, Pl. 10, Fig. 15.

Large rounded hypocone; several large chloroplasts; cingulum tilted downward. **Fig. 30**

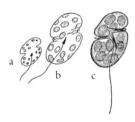

Figure 30. *Gyrodinium pusillum.* (a) (Schilling 1891a, Pl. 10, Fig. 15) (b) (Conrad 1926, Taf 2, Fig. 31) (c) Distinct sulcus and offset cingulum (Thompson 1947, Pl. 1, Fig. 23).

EXTERNAL FEATURES: Elliptical cell; with dorsoventral compression; epicone smaller than hypocone; cingulum displaced one and a half times the cingulum width; sulcus a narrow slit into the epicone, extends into hypocone without reaching antapex.

INTERNAL FEATURES: Eyespot in sulcus beneath cingulum; plastids golden.

SIZE: 25–32 µm long by 18–20 µm wide, 14–15 µm thick (Thompson 1950).

ECOLOGICAL NOTES: Collected in a river estuary in January.

LOCATION REFERENCE: ON (Duthie and Socha 1976 as *Gymnodinium pusillum*), MD (Thompson 1947).

Haidadinium

Buckland-Nicks, Reimchen, and Garbary 1997, 1937, Fig. 6.

TYPE SPECIES: *Haidadinium ichthyophilum* Buckland-Nicks, Reimchen, and Garbary 1997, 1937, Fig. 6.

Epizoic on fish in acidic freshwater.

Assimilative stage round; photosynthetic with golden plastids; epizoic on fish; protoplast with rigid fenestrated matrix. Host fish forms layers of gelatinous, epithelial coating around cells; fish does not seem harmed (Reimchen and Buckland-Nicks 1990). One species reported.

Haidadinium ichthyophilum

Buckland-Nicks, Reimchen, and Garbary 1997, 1937, Fig. 6.

Golden gelatinous film on stickleback. **Fig. 31**

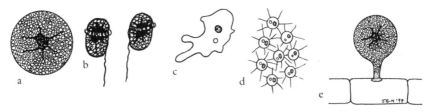

Figure 31. *Haidadinium ichthyophilum.* (a) Cyst on fish (b) Swarmers (c) Lobose amoeba (d) Rhizopodial amoebae (e) Attached to filamentous algae (Buckland-Nicks et al. 1997, Fig. 6 A, C, G, H, I).

EXTERNAL FEATURES: Cell round.

INTERNAL FEATURES: Assimilative cell golden (greenish to brown); nucleus central; protoplast filled with hard, brittle, fenestrated matrix through which run strands of cytoplasm; pyrenoid associated with chloroplast; cytosol contains starch grains and oil bodies.

SIZE: 20–150 µm diameter.

ECOLOGICAL NOTES: Assimilative cells divide repeatedly without daughter cells enlarging; zoospores seen; no evidence of assimilative cell being a trophont (feeding stage); assimilative cell seen to release amoeboid cells (Buckland-Nicks et al. 1990, Buckland-Nicks and Reimchen 1995). Scraped from gelatinous film on stickleback in acidic (pH 4.1) pond.

LOCATION REFERENCE: BC (Buckland-Nicks et al. 1997).

Hypnodinium

Klebs 1912, 443.

TYPE SPECIES: *Hypnodinium sphaericum* Klebs 1912, 399–401, 443, Fig. 8 (402), Taf. 10, 1a, b.

Spherical, planktonic, immobile, golden cell.

Floating, nonmotile unicell; spherical; protoplast shows gymnodinioid organization (cingulum and sulcus, parietal plastids, eyespot) but there are no flagella; envelope closely appressed. This genus may be the cyst stage of a motile species, may be part of the life history of *Cystodinium* (zygote?) (Pfiester and Lynch 1980), or may be valid since reproduction via aplanospores was observed (Klebs 1912); see "Taxonomy" in the introduction.

Key to species of *Hypnodinium*

1a. Plastids in rosettes, cells 64–90 µm diameter .... *H. sphaericum*
1b. Plastids single or in pairs, cells 34–48 µm diameter .... *H. monodisparatum*

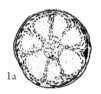   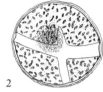

Plate 8. Species of *Hypnodinium*. 1. *H. sphaericum*. (a) Optical daisy in cross section (b) Cell with envelope, ventral view (Klebs 1912, Figs. 8A, B) (c) Rosettes of chloroplasts on cell periphery; note eyespot (Thompson 1949, Fig. 15). 2. *H. monodisparatum* ventral view with scattered single chloroplasts (Hilgert 1976, Pl. 2, Fig. 5, with permission).

Hypnodinium monodisparatum

Hilgert 1976, 93–94, Pl. 2, Fig. 5.

Plastids single or in pairs, cells 34–48 μm diameter. **PL8(2)**

EXTERNAL FEATURES: Spherical nonmotile cell with thick wall; median cingulum barely indented, without displacement; sulcus slightly entering epicone, with parallel sides to antapex.

INTERNAL FEATURES: Numerous golden, lens-shaped plastids scattered as singles or in pairs; may have diffuse red body near center.

SIZE: 34–48 μm diameter, 36–42 μm long
(Hilgert 1976).

ECOLOGICAL NOTES: Collected from the plankton.

LOCATION REFERENCE: MT (Hilgert 1976).

Hypnodinium sphaericum

Klebs 1912, 399–401, 443, Fig. 8 (402), Taf. 10, 1a, b.

Plastids in rosettes, cells 64–90 μm diameter, optical daisy in cross section. **PL8(1); CP3(4)**

EXTERNAL FEATURES: Round cell; smooth wall; cingulum and sulcus visible.

INTERNAL FEATURES: Plastids appear as a parietal network of rosettes/stars on cell periphery; red pigment spot; centrally located nucleus with cytoplasmic strands radiating from it to the periphery; optical section through middle of cell gives appearance of a daisy.

SIZE: 64–90 (up to 133) µm diameter.

ECOLOGICAL NOTES: Planktonic, collected from brown-water pond, November (NC, Whitford and Schumacher 1969), swampy ponds July to September (MD, Thompson 1949).

Katodinium

Fott 1957, 287.

SYNONYM: *Massartia* Conrad 1926, 70–72, Pl. 1 Part 1.

TYPE SPECIES: *Katodinium nieuportensis* (Conrad) Loeblich Jr. and Loeblich III 1966, 37.

Cingulum divides athecate cells into large epicone, small hypocone.

Freshwater or marine; athecate cell with submedian cingulum dividing cell into large epitheca, small hypotheca; cell often mushroom-shaped; often dorsoventrally compressed. Sulcus may enter epicone, may continue into hypocone. Nutrition varies, some with plastids.

Species distinguished by sulcal features, presence or absence of plastids, overall shape, presence or absence of eyespot. There are 23–27 species in one monograph (Christen 1961), 13 in Popovský and Pfiester (1990). Loeblich III (1965) considered that Fott had invalidly transferred species to *Katodinium* and so Loeblich III and not Fott is the correct secondary authority. Thin plates have been found in some species (Jørgensen et al. 2004). Calado (2011) transferred heterotrophic species with an eyespot to *Opisthoaulax*.

Key to species of *Katodinium*

1a. Cell photosynthetic, yellow to golden 2
1b. Cell colorless or grainy, lacking eyespot 4
2a. Attached, in streams .... *K. auratum*
2b. Planktonic, no eyespot 3
3a. Antapex flattened, golden-yellow plastids around nucleus .... *K. planum*
3b. Antapex rounded, several brownish plastids .... *K. bohemicum*
4a. Epicone about same diameter as hypocone .... *K. glandulatum*
4b. Hypocone distinctly smaller than epicone 5
5a. Epicone rounded, hypocone rounded .... *K. fungiforme*
5b. Epicone conical, hypocone lobed .... *K. spirodinioides*

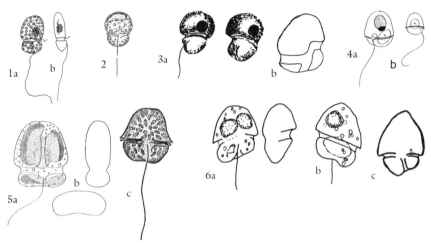

Plate 9a. Species of *Katodinium*. 1. *K. aureatum*. (a) Ventral view with nucleus in epicone, numerous plastids (b) Side view showing dorsoventral compression (Bursa 1970, Figs. 9, 11). 2. *K. bohemicum* ventral view (Thompson 1950, Fig. 13). 3. *K. fungiforme*. (a) Two views (Anissimova 1926, Figs. 9, 10) (b) Ventral view (drawn from micrographs in Spero 1979, Fig. 6, with permission). 4. *K. glanulatum;* note central position of nucleus (Herdman 1924, Figs. 30, 31 as *Gymnodinium glandula*). 5. *K. planum*. (a) Ventral view showing two large chloroplasts (b) Side and top views showing strong dorsoventral compression (Fott 1938, Figs. 2a,b,c) (c) (Thompson 1947, Plate 1, Fig. 20 as *Massartia musei*). 6. *K. spirodinioides*. (a) Ventral (b) Dorsal (sketches of a and b based on Christen 1961, Fig. 17) (c) Ventral (Carty 2003, Fig. 2G).

Katodinium auratum

Bursa 1970, 148–150, Figs. 9–14.

Golden, in streams. **PL9a(1)**

EXTERNAL FEATURES: Elliptical cell; smoothly rounded epicone and hypocone; strongly dorsoventrally compressed; deeply incised cingulum divides cell into two-thirds epicone, one-third hypocone; sulcus brief, not entering epicone and scarcely into hypocone.

INTERNAL FEATURES: Protoplast with numerous golden plastids; nucleus in epicone.

SIZE: 24–26 µm long, 15–16.5 µm wide (Bursa 1970).

ECOLOGICAL NOTES: Collected (rare) in fast-moving stream, attached along ventral surface to sand grains and plants.

LOCATION REFERENCE: CO (Bursa 1970).

Katodinium bohemicum

(Fott) Litvinenko 1977. [authority cited in Popovský and Pfiester 1990; I was unable to confirm as no 1977 reference is given in their literature cited]

BASIONYM: *Gymnodinium bohemicum* Fott 1938, 102, Fig. 1.

Small, photosynthetic, no eyespot. **PL9a(2)**

EXTERNAL FEATURES: Ovate cell; epicone and hypocone broadly rounded; cingulum slightly submedian, without displacement; sulcus indistinct.

INTERNAL FEATURES: Photosynthetic with 2–8 golden (pale yellow green early in the year) parietal plastids (Thompson 1950); refractive granules in cytoplasm; no eyespot.

SIZE: 10–16 µm long; epicone 8–10 µm wide, hypocone 6–7 µm wide.

ECOLOGICAL NOTES: Collected in surface tow from pond, March.

LOCATION REFERENCE: KS (Thompson 1950), TN (Johansen et al. 2007).

Katodinium fungiforme

(Anissimova) Loeblich III 1965, 16.

BASIONYM: *Gymnodinium fungiforme* Anissimova 1926, 191, Figs. 9, 10.

Cell mushroom-shaped, nonphotosynthetic, no eyespot. **PL9a(3)**

EXTERNAL FEATURES: Cingulum almost median but epicone larger than hypocone; sulcus in hypocone only, weak.

INTERNAL FEATURES: Nucleus in epicone; no chloroplasts or eyespot; feeds with peduncle.

SIZE: 4–15 µm long by 5–12 µm wide.

ECOLOGICAL NOTES: Species was originally described from marine samples. The life cycle was attuned to feeding behavior; sexual and asexual reproduction occurred when food was available; resting cells were produced when food was depleted (Spero 1979).

LOCATION REFERENCE: OH (Klarer et al. 2000).

Katodinium glandulatum

(Herdman) Loeblich III 1965, 16.

BASIONYM: *Gymnodinium glandula* Herdman 1924, 81, Figs. 30, 31.

Cell ovoid, nonphotosynthetic, no eyespot. **PL9a(4)**

EXTERNAL FEATURES: Roundish cell divided by cingulum into about two-thirds epicone, one-third hypocone. The epicone is described as having an extended apical point that is flattened toward the back (Herdman 1924) but this was not illustrated; cingulum deep and without displacement; sulcus may enter epicone and descends to the left in the hypocone.

INTERNAL FEATURES: Nucleus central; no chloroplasts; may have greenish refractive granules in cytoplasm and red or yellow accumulation bodies (Herdman 1924).

SIZE: 20–35 µm long.

ECOLOGICAL NOTES: Original paper was concerned with (marine) dinoflagellates discoloring sand (Herdman 1924).

LOCATION REFERENCE: BC (Stein and Borden 1979).

Katodinium planum

(Fott) Loeblich III 1965, 16.

SYNONYM: *Massartia plana* Fott 1938, 103, Fig. 2.

Yellow-gold chloroplasts; hypocone flattened. **PL9a(5)**

EXTERNAL FEATURES: Button mushroom–shaped; epicone somewhat wider than hypocone, which is flattened; sulcus extends deeply into epicone; cell strongly dorsoventrally compressed.

INTERNAL FEATURES: Nucleus central; two large chloroplasts present; no eyespot.

SIZE: 16–20 µm by 15–18 µm.

ECOLOGICAL NOTES: Swims with a jerking movement.

LOCATION REFERENCE: BC (Stein and Borden 1979), MD (Thompson 1947 as *K. musei*), TN (Johansen et al. 2007).

Katodinium spirodinioides

Christen 1961, 332, Fig. 17.

Nonphotosynthetic; conical epicone; no eyespot. **PL9a(6)**

EXTERNAL FEATURES: Small cell; strongly dorsoventrally flattened; larger epicone overhangs smaller hypocone; hypocone distinctly asymmetrical, one side with a larger lobe than the other; cingulum angled across dorsal surface, weakly continued to ventral face; sulcus in hypocone to antapex, weak.

INTERNAL FEATURES: Nonphotosynthetic with a clear to gray and grainy-looking cytoplasm; nucleus in hypocone; stigma lacking.

SIZE: 12–14 µm long by 9–10 µm wide (Christen 1961).

ECOLOGICAL NOTES: Collected in July from a small pond.

LOCATION REFERENCE: OH (VF 970808-1).

The following species of *Katodinium* have also been reported:

Katodinium rotundatum

Reported as blooming from winter-stratified Potomac River (MD, USA) estuary (Cohen 1985).

Katodinium rotundatum

Later found to have plates and scales and was transferred to *Heterocapsa* (Hansen 1995).

Opisthoaulax

Calado 2011, 646–647.

TYPE SPECIES: *Opisthoaulax vorticella* (Stein) Calado 2011, 647.

SYNONYM: *Katodinium* Fott 1957, 287.

Colorless, athecate, phagotrophic cells with epicone distinctly larger than hypocone; red eyespot in sulcus. Resting cyst with pointed ends, eyespot, and several lumpy projections.

Key to species of *Opisthoaulax*

1a. Hypocone flat or bilobed .... *O. musei*
1b. Hypocone rounded 2
2a. Epicone 7–12 µm wide .... *O. tetragonops*
2b. Hypocone 18–30 µm wide .... *O. vorticella*

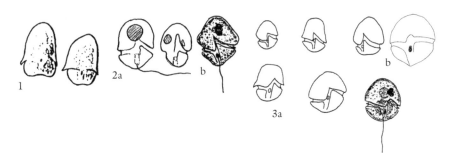

Plate 9b. Species of *Opisthoaulax*. 1. *O. musei;* note overlapping margins of epicone (Pouchet 1887 via Lemmermann 1910, Figs. 13, 14). 2. *O. tetragonops.* (a) Two cells, ventral view (Harris 1940, Figs. 4G,H) (b) Ventral view (Thompson 1950, Fig. 100). 3. *O. vorticella.* (a) Ventral views (Thompson 1950, Figs. 94–99) (b) Ventral (orig.).

Opisthoaulax musei

(Danysz in Pouchet) Calado 2011, 647.

BASIONYM: *Gymnodinium musaei* Danysz in Pouchet 1887, 89, 104–105, 112, Pl. 10, Fig. 6.

SYNONYMS: *Massartia musei* (Danysz) Schiller 1933, 438; *Katodinium musaei* (Danysz in Pouchet) Loeblich III 1965, 16.

Nonphotosynthetic, flaring epicone, eyespot appears as double rod. **PL9b(1)**

EXTERNAL FEATURES: Oval cell; dorsoventrally flattened; posterior flattened to slightly lobed; cingulum below median without displacement; lower margin of epicone flaring out over cingulum; sulcus in hypocone to antapex.

INTERNAL FEATURES: Distinctive eyespot described as two red rods oriented parallel to each other (now considered the sides of a trough-shaped structure) near the surface in the sulcus (Pouchet 1887); the protoplast is described as clear and containing spherical green globules, there is no mention of plastids (Pouchet 1887).

SIZE: 19–20 μm long by 18 μm wide, 11 μm thick.

LOCATION REFERENCE: AR (Meyer 1969), MN (Meyer and Brook 1969).

COMMENTS: Thompson (1947) collected something he called *Massartia musei* with the general shape of the Pouchet cells but lacking the distinctive eyespot, and with plastids (I have placed it in *K. planum*, though the chloroplasts do not match); Meyer and Brook (1969) reported the species from a *Sphagnum* bog without illustration.

Opisthoaulax tetragonops

(Harris) Calado 2011, 647.

BASIONYM: *Massartia tetragonops* Harris 1940, 15, Fig. 4G–L.

SYNONYM: *Katodinium tetragonops* (Harris) Loeblich III 1965, 16.

Nonphotosynthetic, small; with eyespot. **PL9b(2)**

EXTERNAL FEATURES: Metabolically plastic from oval to rhomboidal; indented cingulum descending on left side of cell, crosses dorsal surface, offset one cingulum width on right side of cell; sulcus narrow with parallel sides, extending into epicone (origin of cingulum) and almost to antapex, gives backward-Z configuration to cingulum-sulcal grooves; hypocone rounded; cell dorsoventrally compressed.

INTERNAL FEATURES: Square red eyespot in sulcus, chromatophores lacking; cytoplasm colorless and granular.

SIZE: 10–15 μm long by 7–12 μm wide (Harris 1940, Thompson 1950).

ECOLOGICAL NOTES: Active swimmer.

LOCATION REFERENCE: KS (Thompson 1950).

Opisthoaulax vorticella

(Stein) Calado 2011, 647.

BASIONYM: *Gymnodinium vorticella* Stein 1883, Pl. 3, Figs. 1–4.

SYNONYMS: *Massartia vorticella* (Stein) Schiller 1933, 441; *Katodinium vorticellum* (Stein) Loeblich III 1965, 16.

Nonphotosynthetic; with eyespot; cell ovoid. **PL9b(3); CP3(3)**

EXTERNAL FEATURES: Ovoid cell; indented cingulum originates at the sulcus in approximately midcell, descends on the left side of cell across the dorsal surface, and appears on the right side of the cell at least one cingulum width displaced; sulcus (cingulum origin) does not enter epicone and has parallel sides in the hypocone but does not reach antapex.

INTERNAL FEATURES: May contain ingested algae, red oil bodies; red eyespot in sulcus (Thompson 1950).

SIZE: 21–33 µm long by 18–30 µm wide.

Piscinoodinium

Lom 1981, 8.

TYPE SPECIES: *Piscinoodinium pillulare* (Schäperclaus) Lom 1981, 9.

BASIONYM: *Oodinium pillularis* Schäperclaus 1951, 169–171 (specific epithet only), 1954 (combination).

Freshwater ectoparasite of fish gills and skin.

Trophont (feeding) stage on fish followed by encystment, production of gymnodinioid swarmers that infect fish.

Piscinoodinium limneticum

(Jacobs) Lom 1981, 9.

BASIONYM: *Oodinium limneticum* Jacobs 1946, 10–12, Pl. 1–2, Figs. 1–15.
Yellowish dusting on freshwater fish. **Fig. 32**

EXTERNAL FEATURES: Cells oval to pyriform; smooth, thin walled.

INTERNAL FEATURES: Variably shaped, light green plastids; nucleus central, about 6 µm diameter with obvious granules; no eyespot; no flagella.

SIZE: Cells initially 12–20 µm long by 7.5–13 µm wide.

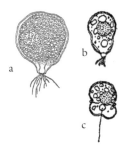

Figure 32. *Piscinoodinium limneticum.* (a) Almost mature parasite (b) Infectious swarmer (c) Swarmer shortly after attachment (sketch based on Jacobs 1946, Pl. 1, Figs. 2, 8, 9).

ECOLOGICAL NOTES: Infection appears as yellowish (on dark fish) to whitish (on light fish) dusting on all parts of fish body, especially scale edges (Jacobs 1946). Cell body attached to fish by stalk with rhizoid-like processes that grow between host cells. As cells feed they grow in length: 28 µm after one day, 38 µm after two, 50 µm after three, and 60 µm after four days, the width being two-thirds the length. Further growth in width produces a roundish cell up to 80 µm in diameter, 96 µm long with the stalk. Large cells contain starch and numerous olive plastids. Cyst formation occurs when the stalk and rhizoids are retracted and the pore through which they were extended is sealed. Repeated cell division eventually produces gymnodinioid swarmers 10–19 µm long with a distinct median cingulum that divides the cell into a larger rounded epicone and smaller bilobed hypocone. The sulcus reaches the antapex; there is no eyespot. If this infective stage does not contact a fish, it forms a cyst. If it does contact a fish, a peduncle emerges from the sulcus and penetrates host tissue (Jacobs 1946). Causes fish death in aquaria and can parasitize zebra fish, guppy, mollie, gourami, platy, and swordtail (Jacobs 1946).

Lom (1981) in reviewing parasitic dinoflagellates erected several genera for the fish parasites to distinguish them from *Oodinium*, a parasite on tunicates. Work on the type species (*P. pillulare*) suggests the taxon is primarily photosynthetic (well-developed plastids and starch grains) with perhaps osmotrophic uptake of host metabolites (Lom and Schubert 1983). Molecular analysis of *Piscinoodinium* infecting killifish placed the genus in order Gymnodiniales (Levy et al. 2007).

LOCATION REFERENCE: Mexico (Ortega 1984).

Prosoaulax

Calado and Moestrup 2005, 113.

TYPE SPECIES: *Prosoaulax lacustris* (F. Stein) Calado and Moestrup 2005, 113.

Freshwater athecate cells with small epicone and larger hypocone.

Genus established for the freshwater species formerly in the genus *Amphidinium*. Cells with epicone less than or equal to one-third total length; length about the same as width; cingulum without displacement; may have an eyespot; yellowish plastids, if present, may be kleptoplastids (Calado and Moestrup 2005).

Prosoaulax lacustris

(F. Stein) Calado and Moestrup 2005, 113.

BASIONYM: *Amphidinium lacustre* Stein 1883, 15, Taf. 17, Figs. 21–30.

Small epicone, heart-shaped hypocone, heterotrophic. **Fig. 33**

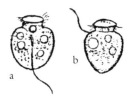

Figure 33. *Prosoaulax lacustris.* (a) Ventral (b) Dorsal (Stein 1883, Taf 17, Figs. 21, 22; note transverse flagellum depicted as cilia).

EXTERNAL FEATURES: Epicone oval, centered over, but much smaller than hypocone; sulcus does not enter epicone and is wide at the junction with the cingulum, narrowing and almost reaching the antapex, giving the hypocone a heart-shaped appearance.

INTERNAL FEATURES: No eyespot; may retain the plastids of prey (Calado and Moestrup 2005).

SIZE: 23 µm long by 18 µm wide.

ECOLOGICAL NOTES: Heterotrophic feeding on cryptomonads and other flagellates (Calado et al. 1998, Jørgensen et al. 2004).

LOCATION REFERENCE: BC (Wailes 1928a, 1934).

Pseudoactiniscus

Bursa 1969, 414–416, Fig. 19.

TYPE SPECIES: *Pseudoactiniscus apentasterias* Bursa 1969, 414–416, Fig. 19. Freshwater cell with shallow cingulum, sulcus extends to antapex and as a groove to the apex.

Spherical cell; shallow, median cingulum without displacement; sulcus in hypocone with parallel sides to antapex, extending in epicone to apical dot; fragile cell envelope with rows of punctate inclusions; nucleus central; protoplast contains vacuoles but not pentasters (Bursa 1969).

Pseudoactiniscus apentasterias

Bursa 1969, 414–416, Fig. 19.

Without pentasters, in cold Arctic lake. **Fig. 34**

Figure 34. *Pseudoactiniscus apentasterias* (redrawn from Bursa 1969, Fig. 19).

EXTERNAL FEATURES: Spherical cell; shallow median cingulum without or with slight displacement; sulcus in hypocone with parallel sides, reaching antapex, continuing in epicone to apical dot. Reproduction by binary fission.

INTERNAL FEATURES: Protoplast with lipid drops, hematochrome grains, green-yellow chromatophores, vacuoles; centrally located nucleus.

SIZE: 26–53 µm diameter.

ECOLOGICAL NOTES: Ingestion of *Cryptomonas* and *Gymnodinium;* found in Keyhole Lake, NU, surface temperature 1–11 °C June to September, small bloom of 1520 cells/L (Bursa 1969).

LOCATION REFERENCE: Keyhole Lake, NU (Bursa 1969).

COMMENTS: Spherical cysts listed in the species description and mentioned as a distinguishing difference from *Actiniscus canadensis* were not illustrated. Because many of the cells of *Actiniscus canadensis* lacked pentasters and *Pseudoactiniscus* lacks pentasters, the differences between these two taxa are small. Further, the illustration (Fig. 19 in Bursa 1969) defining *Pseudoactiniscus* is a lighter version, but identical to Fig. 21, supposedly of *Actiniscus canadensis.*

Rufusiella

Loeblich III. In Christensen 1978, 68.

BASIONYM: *Haematococcus insignis* Hassall 1845, subgenus *Ouracoccus,* 324, Pl. 80, 6a,b.

SYNONYMS: *Ouracoccus* Lindley 1846; *Urococcus* Kutzing 1849, invalid for *Ouracoccus.*

TYPE SPECIES: *Rufusiella insignis* (Hassell) Loeblich III 1967, 231.

Round cells in concentrically laminated sheath, found on wet sandstone and drip walls.

Round cells in laminated sheath; variously pigmented (green, red, golden); scraped from sandstone kept moist by freshwater. This taxon was initially identified as a *Haematococcus* (Chlorophyta) which is red; *Haematococcus* subgenus *Ouracoccus* was then raised to genus level. Its identity as a dinoflagellate was suggested to Smith (1950) by Thompson, who also placed it with the dinoflagellates (Thompson 1959). It is described as imparting a reddish tinge in the Arctic intertidal (Fensome et al. 1993). Since *Urococcus* (an invalid name for *Ouracoccus*) was initially listed as a green alga, its inclusion on species lists will not be considered the dinoflagellate unless it is listed under Dinophyceae. The description following is based on the work of Richards (1962), a student of Thompson's.

Rufusiella insignis

(Hassell) Loeblich III 1967.

BASIONYM: *Haematococcus insignis* Hassall 1845, subgenus *Ouracoccus,* 324, Pl. 80, Fig. 6a, b.

Round cells in concentrically laminated sheath, found on sandstone and drip walls. **Fig. 35; CP3(5)**

EXTERNAL FEATURES: The sheath is made of one to many concentric lamellae; division products (2,4, sometimes 8) remain in parental sheath;

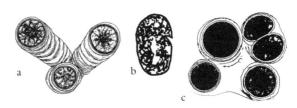

Figure 35. *Rufusiella insignis.* (a) Stalklike laminated sheaths (b) Zoospore (c) Spherical cells and recently divided cells (Richards 1962, Figs. 3, 13, 9).

accumulated sheathing material may form stalklike structures with the cells terminal. Zoospores thecate (4′, 2–7a, 7″, 7‴, 1–2p, 2⁗), golden, *Hemidinium*-like, 26–46 μm long and 15–30 μm wide, no eyespot (Richards 1962).

INTERNAL FEATURES: Olive to brown or orange; cytosol with starch grains and elliptical, radially arranged plastids.

SIZE: Mature cells 30–83 μm diameter (41–100 μm including sheath); after division young cells are 29–40 μm long by 25–29 μm wide.

ECOLOGICAL NOTES: Cells and colonies are indistinguishable from *Gloeo-dinium montanum;* both grow in acidic habitats, and they may be synony-mous, but for now the difference in habitat will be considered sufficient to keep them as separate genera. Collected from moist sandstone in Kansas and from drip walls in Great Smoky Mountains National Park, TN.

Stylodinium

Klebs 1912, 410–411, 445.

Roundish cell on stipe, often attached to *Oedogonium.*

Roundish (spherical, oval, obovate, squarish) solitary cell attached by stipe (stalk) to a filamentous green alga. Stipe varies in length from almost sessile to cell length or longer, has a disk-shaped holdfast at the base and swelling at the junction with the cell. Parasitic genus, the plastids are not its own and range in color from bright green to orange and brown depending on stage of metabolism.

Species are separated on cell shape, stipe length and features of the plas-tids (Bourrelly 1970) though stipe length is variable (Thompson 1949, Popovský and Pfiester 1990), eight species are listed in Starmach (1974), one, *S. globosum,* in Popovský and Pfiester (1990).

Key to species of *Stylodinium*

1a. Cell squarish with long stipe .... *S. longipes*
1b. Cell round, stipe length variable .... *S. globosum*

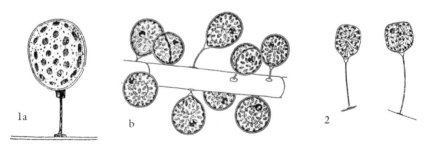

Plate 10. Species of *Stylodinium.* 1. *S. globosum.* (a) Single stalked cell (Klebs 1912, Fig. 12 A) (b) Several cells; note rounded shape (Thompson 1949, Fig. 25). 2. *S. longipes.* Note squarish shape, long stipe (Thompson 1949, Figs. 30, 32).

Stylodinium globosum

Klebs 1912, 445.

SYNONYM: *Stylodinium sphaera* Pascher 1944a, 393–394, Fig. 15.

Globose cell, attached to *Oedogonium* with a stipe about the length of the cell diameter. **PL10(1); CP3(6,7)**

EXTERNAL FEATURES: Cell globose; attached by stipe, short to 15 µm long, with distinct disk-shaped holdfast and swelling at the junction with the cell body.

INTERNAL FEATURES: Thompson (1949) described parietal discoid chromatophores, but I have seen cells olive, golden, and orange in color with red or brown accumulation bodies; reproduction by formation of daughter protoplasts.

SIZE: Cell diameter 21–35 µm.

ECOLOGICAL NOTES: The *Oedogonium* cell immediately beneath the *Stylodinium* will almost always be empty (parasitized). Also reported on *Utricularia* from Panama (Prescott 1951a). Pfiester and Popovský (1979) studied this species (as *Stylodinium sphaera*) for details of its parasitism.

Stylodinium longipes

Thompson 1949, 308, Figs. 30–34.

Squarish cell, long stipe attached to *Oedogonium*. **PL10(2)**

EXTERNAL FEATURES: Cell squarish; stipe with holdfast disk and swelling at junction with cell.

INTERNAL FEATURES: Nucleus large; plastids golden brown, strap-shaped, and radially arranged; red oil globule present (Thompson 1949). Division of protoplast into two.

SIZE: Cell 15–38 µm long by 12–28 µm wide, 8–22 µm thick; stipe 6–36 µm long.

ECOLOGICAL NOTES: Collected August–September from swampy pool attached to *Oedogonium*.

LOCATION REFERENCE: MD (Thompson 1947).

Tetradinium

Klebs 1912, 408–410, 444.

TYPE SPECIES: *Tetradinium javanicum* Klebs 1912, 444, Fig. 11, p. 408.

Attached, tetrahedral, nonmotile, golden cell with spines.

Cells triangular or squarish from above, tapering to a short stipe with holdfast disk; corners with one or two short spines. Plastids numerous, small, golden-brown disks; nucleus central; small red oil globule present. Reproduction by two zoospores.

Key to species of *Tetradinium*

1a. Cell with one downward-pointing spine at each corner .... *T. simplex*
1b. Cell with two spines at each corner .... *T. javanicum*

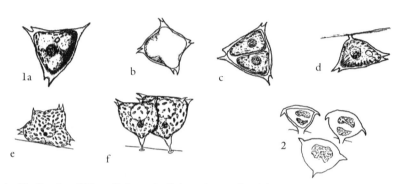

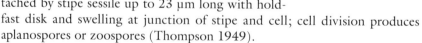

Plate 11. Species of *Tetradinium*. 1. *T. javanicum* cells with usually two spines at corners. (a) Cell triangular in top view (b) Cell quadrangular in top view (c) Cell dividing (d) Side view showing attachment (Klebs 1912, Figs. 11A,C,D,G) (e) Four-lobed cell (f) Attached cells showing stipe and holdfast (Thompson 1949, Fig. 20A). 2. *T. simplex* cells with single spine at corners (sketch based on Prescott et al. 1949, Figs. 13, 14).

Tetradinium javanicum

Klebs 1912, 444, Fig. 11.

SYNONYMS: *Tetradinium intermedium* Geitler 1928a, 2–4; *Tetradinium minus* Pascher 1927, 22, Fig 21.

Attached or planktonic, golden tetrahedral cell with two spines at each corner. **PL11(1); Figs. 36–38; CP3(8,9)**

EXTERNAL FEATURES: Cell shaped like inverted pyramid; sides convex; spines at corners splayed; attached by stipe sessile up to 23 µm long with holdfast disk and swelling at junction of stipe and cell; cell division produces aplanospores or zoospores (Thompson 1949).

INTERNAL FEATURES: Chromatophores golden brown, parietal or scattered; nucleus central; large central vacuole.

SIZE: Cell 20–73 µm diameter.

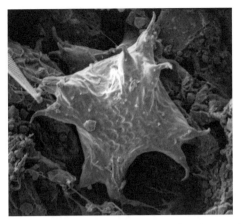

Figure 36. *Tetradinium javanicum.* Note spine extensions in several planes (SEM).

Figure 37. Body and spines in focus.

Figure 38. Stipe in focus (LM).

ECOLOGICAL NOTES: Attached to *Fragilaria* or filamentous algae like *Oedogonium,* pond weeds *Myriophyllum* and *Potamogeton,* or empty arthropod exoskeletons (insect or zooplankton), or may be free floating (Thompson 1949). Illustrations of *T. intermedium* are not readily distinguished from *T. javanicum.*

Tetradinium simplex

Prescott in Prescott et al. 1949, Pl. 2, Figs. 13, 14.

Attached triangular cell with single, downward-pointing spine at each corner. **PL11(2)**

EXTERNAL FEATURES: Cell triangular; sides convex; angles with a single, downward-pointing spine; attached but sessile.

SIZE: 12–25 µm diameter.

ECOLOGICAL NOTES: Collected on filamentous algae (Prescott et al. 1949).

LOCATION REFERENCE: MI (Prescott et al. 1949).

THECATE/ARMORED TAXA

Amphidiniopsis

Wołoszyńska 1928, 174 (Polish) or 256 (German).

TYPE SPECIES: *A. kofoidii* Wołoszyńska 1928, 256, Taf. 7, Figs. 1–17. Sand dwelling, amphidinioid-shaped.

Thecate, *Amphidinium*-like (distinct cingulum divides cell into small anterior and large posterior sections), nonphotosynthetic. Most species marine.

Amphidiniopsis sibbaldii

Nicholls 1998, 337, Fig. 20; Nicholls 1999.

Freshwater, sand dwelling, thecate, amphidinioid. **Fig. 39**

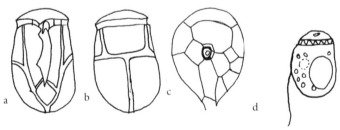

Figure 39. *Amphidiniopsis sibbaldii.* (a, b, c) Plates in ventral, dorsal, and apical views, respectively (d) Internal features visible with the light microscope (sketch based on Nicholls 1998, Figs. 19, 20 A, B, D).

EXTERNAL FEATURES: Cell laterally compressed, cingulum divides cell into about one-seventh epitheca, six-sevenths hypotheca; plate pattern: pp, 4′, 3a,

7″, 5C, 3S, 5‴, 2⁗ with plates 2′ and 3′ diminutive and 3a large. This species is distinguished by the presence of a cutlass-shaped sulcus and large antapical plates (Nicholls 1998).

INTERNAL FEATURES: No plastids, nucleus in dorsal hypotheca, accumulation bodies present, no eyespot.

SIZE: Cells 33–45 µm long by 19–29 µm wide, 24–35 µm deep.

ECOLOGICAL NOTES: Found in depression in sandy shoreline, water temperature 9 °C (later kept in lab at 16–19 °C) (Nicholls 1998).

LOCATION REFERENCE: ON (Nicholls 1998).

Borghiella

Moestrup, Hansen, and Daugbjerg 2008, 56–57.

TYPE SPECIES: *Borghiella dodgei* Moestrup, Hansen, and Daugbjerg 2008, 57, Fig. 7.

Roundish photosynthetic cell with eyespot, with numerous, thin, hexagonal plates.

This genus was extracted from *Woloszynskia* based on *W. tenuissima* for species with numerous, thin, usually hexagonal plates; an apical furrow; an eyespot as part of a chloroplast; and a round cyst (Moestrup et al. 2008).

Borghiella tenuissima

(Lauterborn) Moestrup, Hansen, and Daugbjerg 2008, 74.

BASIONYM: *Gymnodinium tenuissimum* Lauterborn 1894 (description only), 1899, 388, Taf. 17, Fig. 26.

SYNONYMS: *Gymnodinium tenuissimum* Lauterborn as illustrated by Wołoszyńska 1917, Pl. 11, Figs. 7–9, Pl. 12, Figs. 1–4, Pl. 13 M; *Woloszynskia tenuissima* (Lauterborn) Thompson 1950, 290.

Strongly dorsoventrally flattened cell with incomplete cingulum; cryophile.
PL12(1)

EXTERNAL FEATURES: Flat cells, ventral surface concave; cell somewhat ovoid, the epitheca conical, coming to a small point; the hypotheca broadly rounded, may be slightly lobed; cingulum incomplete, not fully reaching sulcus on right side; narrow sulcus not extending into epitheca and not reaching antapex; short apical furrow; theca of many small, thin plates (Moestrup et al. 2008).

INTERNAL FEATURES: Eyespot part of chloroplast (Moestrup et al. 2008); nucleus in hypocone; numerous golden chloroplasts on cell periphery (Crawford et al. 1970).

SIZE: 40 µm long by 33 µm wide (Crawford et al. 1970).

ECOLOGICAL NOTES: Cold-loving species isolated from beneath the ice in Greenland (Moestrup et al. 2008).

LOCATION REFERENCE: KY (Dillard 1974), Greenland (Moestrup et al. 2008).

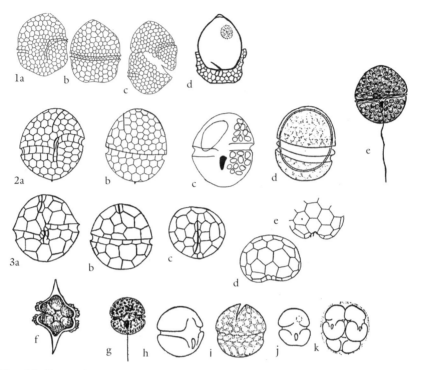

Plate 12. Taxa with numerous small plates. 1. *Borghiella tenuissima*. (a) Ventral view; note incomplete cingulum, theca of numerous small polygonal plates (b) Dorsal view (c) Splitting of theca in sulcal area (d) Cyst, may be round (Wołoszyńska 1917, Pl. 11, Figs. 7–9, Pl. 13 M, as *Gymnodinium tenuissimum*). 2. *Jadwigia applanata*. (a) Ventral; note displaced cingulum and subapical line (b) Dorsal (Wołoszyńska 1917, Taf 11, Fig. 5, Taf 12, Fig. 11 as *Glenodinium neglectum*) (c) Ventral view; note eyespot in sulcus, numerous peripheral chloroplasts (just drawn on left side of cell) (original, assembled from Figures 40, 44, 46, 54 in Lindberg et al. 2005) (d) Dorsal showing separation at cingulum (e) Ventral; note eyespot in sulcus, radially arranged chloroplasts (Thompson 1947, Pl. 1, Figs. 15, 16 as *Gymnodinium neglectum*). 3. *Tovellia glabra*. (a) Ventral (b) Dorsal (c) Apical (d, e) Antapex showing several plates (f) Cyst (Wołoszyńska 1917, Taf 11, Figs. 13, 14 19, Taf 13, Fig. L as *Gymnodinium coronatum*, Taf 11, Figs. 20, 21 as *Gymnodinium coronatum* var. *glabra*) (g) Ventral, showing eyespot and chloroplasts (h, j) Sulcus and eyespot in larger and smaller cells (i) Plate pattern (k) Result of division (Thompson 1950, Figs. 105–109 as *Woloszynskia coronata*).

Ceratium

Schrank 1793, 34–35, Table 3, Fig. 23.

TYPE SPECIES: *Ceratium hirundinella* (O.F.M.) Dujardin 1841, 377.

SYNONYM: *Ceratium tetraceras* Schrank 1793 (see Calado and Larsen 1997).

One apical horn, at least two posterior horns, common.

The most common, most recognized, freshwater dinoflagellate. Plate pattern: 4′, 5″, 5‴, 2⁗; five cingular plates (four in marine species). Apical horn may be straight or curved, with or without an apical pore, and is composed of three or four of the four apical plates. Ventral area is deeply concave; cingulum without displacement; cingular plate sutures aligned with precingular plate sutures. One posterior horn is composed of both antapical plates; a right posterior horn is composed of the fourth and fifth postcingular plates and may be accompanied by an accessory left postcingular horn composed of plates 1‴ and 2‴ (postcingular horns unique to *Ceratium*). There is reticulate ornamentation on thick thecal plates, each areole with a trichocyst pore. Freshwater species have golden plastids, no eyespot. Triangular cysts with processes corresponding to the number of horns are produced. Vegetative cell division occurs by the separation of plates along a specific suture line into two parts that each regenerate the missing section. Slow, graceful swimming.

Ceratium has worldwide distribution in freshwater and there is a vast literature on the ecology of *Ceratium hirundinella*. *Ceratium* is sometimes implicated as a taste and odor or filter-clogging alga (Knappe et al. 2004). A new freshwater species lacking hypothecal horns has been reported from Greece (Temponeras et al. 2000). There are fossil records (Bint 1986) and marine species. A proposal to transfer all marine species to a new genus, *Neoceratium*, has generated some controversy (Gomez et al. 2010, Calado and Huisman 2010, Gomez 2010). Wall and Evitt (1975) discussed relationships among modern freshwater and marine species and fossils.

Key to species of *Ceratium*

1a. Apical horn at angle 2
1b. Apical horn perpendicular to body... 3
2a. Apical horn strongly curved, tapering to a point, no pore .... *C. carolinianum*
2b. Apical horn short, with an apical pore... *C. cornutum*
3a. Two short, blunt posterior horns... *C. brachyceros*
3b. Two or three slender, tapering posterior horns 4
4a. Epitheca smoothly tapering from cingulum, 2 down-pointing posterior horns...
....*C. furcoides*
4b. Epitheca with shoulders, 2 down-pointing posterior horns .... *C. rhomvoides*
4c. Epitheca with shoulders, 2–3 posterior horns, often splayed...........................
.... *C. hirundinella* and forms

Ceratium brachyceros

Daday 1907, 252, Fig. A.

Two short, blunt posterior horns. **PL13(1); Figs. 40, 41**

EXTERNAL FEATURES: Apical pore and pore plate, 4′, 5″, 4S, 5‴, 2⁗; straight apical horn, hypothecal horns are short and blunt, in particular the one postcingular horn, thick reticulate ornamentation. Examination by SEM shows the 4′ plate not reaching

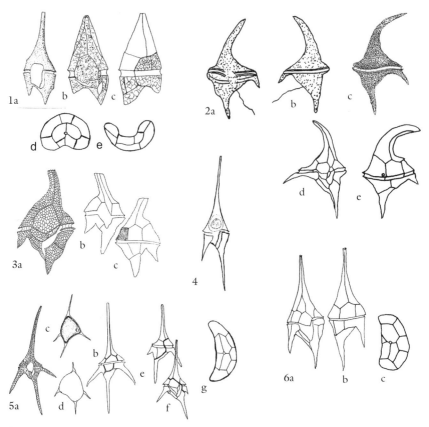

Plate 13. Species of *Ceratium*. 1. *C. brachyceros.* (a) Ventral view (Daday 1905, 252, Fig. A) (b,c) Ventral and dorsal views; note stubby posterior horns (d) Epitheca (e) Hypotheca (orig.). 2. *C. carolinianum.* (a) Ventral (b) Dorsal views (Bailey 1850, Plate 3, Figs. 4, 5) (c) Dorsal view; note pointed apical horn (Whitford and Schumacher 1969, Pl. 59, Fig. 22, with permission) (d) Ventral (e) Dorsal; note pore (Wall and Evitt 1975, text figs. 5A,B, with permission). 3. *C. cornutum.* (a) Ventral view; note angle of apical horn and its blunt apex (Tiffany and Britton 1952, Fig. 994, with permission) (b) Ventral (c) Dorsal views (Prescott 1951b, Pl. 92, Figs. 8, 9). 4. *C. furcoides,* slender elongate cell with gradual taper of apical horn from cingulum (Levander 1894, Taf 2, Fig. 24). 5. *C. hirundinella.* (a) Ventral view with reticulate ornamentation (b) Dorsal view (c, d) Cysts with four processes corresponding to the four horns (Thompson 1947, Pl. 4, Figs. 6–9) (e) Dorsal; division lines for separation (f) Ventral; division line (Entz 1927) (g) Epitheca showing all four apical plates reaching apex. 6. *C. rhomvoides.* (a) Ventral view; note two downward-pointing posterior horns (b) Dorsal view (a and b modified from Hickel 1988, Fig. 19) (c) Epitheca.

the apex and four sulcal plates (Carty 1986). This species is most likely to be confused with *Ceratium rhomvoides,* a larger species with more elongate, slender horns.

SIZE: Small species of *Ceratium,* 65–80 µm long by 33–43 µm wide.

ECOLOGICAL NOTES: Collected from large lakes in summer and fall.

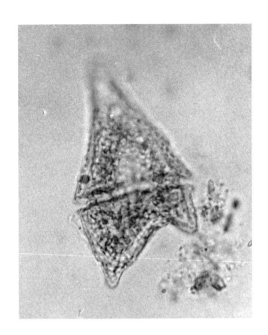

Figure 40. *Ceratium brachyceros.*
Dorsal view (LM).

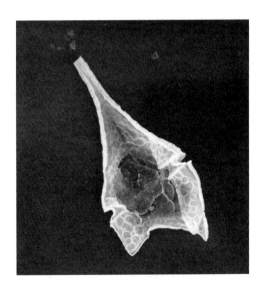

Figure 41. Ventral view with plates
collapsed (SEM).

Ceratium carolinianum

(Bailey) Jörgensen 1911, 14.

BASIONYM: *Peridinium carolinianum* Bailey 1850, 41, Pl. 3, Figs. 4, 5.

SYNONYM: *Ceratium curvirostre* Huitfeldt-Kaas 1900, 1–7.

Strongly curved apical horn tapering to a point. **PL13(2); Figs. 42, 43**

EXTERNAL FEATURES: Large golden cell with curved apical horn tapering to a point (no apical pore), two posterior horns, the posterior horn

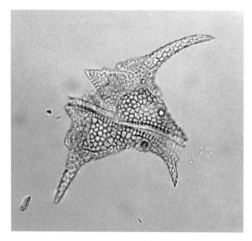

Figure 42. *Ceratium caroliniaum.* Dorsal view (LM).

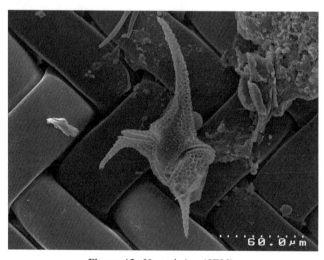

Figure 43. Ventral view (SEM).

composed of antapical plates is directed downward, the single slender posterior horn composed
of postcingular plates is directed outward and
downward, reticulate ornamentation. Wall and
Evitt (1975) and Bourrelly (1970) illustrated the
species with the 1′ plate not reaching the apex.
Pore on the 2″ plate is a slender tube to the interior (Bourrelly and Couté 1976).

SIZE: Cell 160–260 μm long by 65–87 μm wide at
cingulum.

ECOLOGICAL NOTES: First reported from South Carolina, USA "among
the roots of *Lemna* in the 'back-waters' of rice fields" (Bailey 1850), it has
been found in *Sphagnum* bogs and ponds (Prescott 1951b), a *Batrachosper-
mum* pond (Prescott and Croasdale 1937), and oligotrophic lakes in Scan-
dinavia (Morling 1979). It has been collected throughout Europe
(Morling 1979).

Ceratium cornutum

(Ehrenberg) Claparède and Lachmann 1858, Taf. 20,
Fig. 1.

BASIONYM: *Peridinium cornutum* Ehrenberg 1830,
75.

Angled apical horn, with apical pore. **PL13(3)**

EXTERNAL FEATURES: Chunky *Ceratium* with
bent, but straight, bluntly terminated apical horn
with apical pore, two posterior horns, sulcus deeply
concave, cingulum not displaced. The 1′ plate may
not reach the apex (Happach-Kasan 1982, Vyerman and Compère 1991).
The junction of plates 2″ and 3″ near the cingulum contains an area of high
pore density, and micrographs showed a notch in the upper cingular rim at
that location (Bourrelly and Couté 1976). Cells attach to the bottom of
petri dishes by ribbons extruded from the dorsal side between plates 2″ and
3″, near the cingulum (Happach-Kasan 1982). This is the location of the
unusual dorsal pore in *C. carolinianum* and shows an affinity between the
two species and a possible function for the structure.

SIZE: 75–80 μm wide; cysts 80–90 μm in length have been reported
(Morling 1979).

ECOLOGICAL NOTES: Rare in the plankton of softwater lakes (Prescott
1951b).

Ceratium furcoides

(Levander) Langhans 1925, 597.

BASIONYM: *Ceratium hirundinella* var. *furcoides* Levander 1894, 53, Taf. 2, Fig. 24.

Slender, elongate species, two posterior horns, gradual tapering of straight apical horn from cingulum. **PL13(4); Fig. 44**

EXTERNAL FEATURES: Slender, elongate cell, apical horn tapering gradually from cingulum, terminating in an apical pore, the 4′ plate does not reach the apex. The posterior horns are directed perpendicular to the cingulum, cingulum narrow without displacement, ventral face deeply concave, upper and lower portions of the cell not vertical; viewed from the side, the anterior and posterior horns form an obtuse angle.

INTERNAL FEATURES: Chloroplasts yellowish green (Hickel 1988).

SIZE: 150–200 µm long by 30–37 µm wide.

ECOLOGICAL NOTES: It could be a dominant species June to October and occasionally formed a bloom (Hickel 1988). I found a bloom in a small pond in Greenport, NY. Hickel (1988) followed this species through its life cycle and recorded asexual reproduction, sexual reproduction, and production of benthic cysts (hypnozygotes).

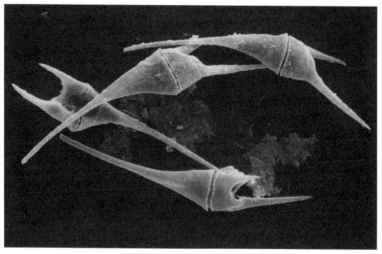

Figure 44. *Ceratium furcoides.* Dorsal and ventral views; note smooth taper of epitheca, no shoulders.

Ceratium hirundinella

(O.F.M.) Dujardin 1841, 377.

BASIONYM: *Bursaria hirundinella* O.F. Müller 1773 (text only), 1786, Pl. 17, Figs. 9–12.

Long straight apical horn, 2–3 posterior horns. **PL13(5); CP5(1–6)**

EXTERNAL FEATURES: Plate pattern: apical pore, pore plate, 4′, 5″, 5C, 5‴, 2‴′; all four apical plates elongated into an anterior horn with apical pore; cingulum median, without displacement; sulcus deeply concave, large Sa plate subtended by an Sp, both with same heavy reticulate ornamentation as body, two smaller sulcal plates (F and O of Entz [1927], now an Sd and Sda) to the left of the Sa and Sp. Reticulate ridges of ornamentation extend into short spines on the horns. Asexual reproduction occurs by separation of the parent cell along specific lines to yield two half cells, each with a portion of the cingular area and parental theca, sexual reproduction occurs by engulfment of a smaller male gamete by the female gamete, resulting in a distinctive three- to four-horned cyst.

INTERNAL FEATURES: Chloroplasts golden to brown; digestive vacuoles containing bacteria, cyanobacteria, and diatoms have been seen (Entz 1927, Dodge and Crawford 1970); food reserves lipid and starch.

ECOLOGICAL NOTES: Cosmopolitan in lakes; it sometimes causes dramatic blooms worldwide (see the introduction).

COMMENTS: Forms (forma) are based on the "Formentypen" morphotypes of Bachman (1911) and have no taxonomic status (Calado and Larsen 1997). They are distinguished primarily on the number, length, and divergences of posterior horns. It has been widely accepted that the forms of *C. hirundinella* are an example of cyclomorphosis. Detailed studies of cyclomorphosis (summarized in Hutchinson 1967) find changes related to temperature, with a spring maximum in horn length and width, and a secondary autumn maximum. Another proposed explanation of the forms is based on nutrition, with larger cells having three posterior horns, the result of greater nutrient (nitrogen) availability (Kimmel and Holt 1988). Another factor affecting shape may be predation; work with mesocosms containing zooplanktivorous fish found *Ceratium hirundinella* with two posterior horns, while enclosures lacking fish had three posterior horns (Hamlaoui et al. 1998). The variety of forms assigned to *C. hirundinella* may be separate species, phenotypic plasticity within a single genome responding to environmental conditions, or a combination of both. Hutchinson (1967) commented that the forms could be species succeeding each other through a season. Evidence for separate taxa includes different sizes, different hypothecal horn morphologies, nonvariability in some populations, and distinctive cysts produced by the different forms (Huber and Nipkow 1922). Some forms have distinctive features, others (*robustum*) are similar to the type. Since the type species, *Ceratium hirundinella* f. *hirundinella* has three posterior horns, it could be argued that forms with two horns are other species (all the other freshwater species of *Ceratium* have two posterior horns).

Key to forms of *Ceratium hirundinella*

1a. Cell with two posterior horns 2
1b. Cell with three posterior horns 3
2a. Cell 350–450 µm long .... *C. h.* form *yuennanense*

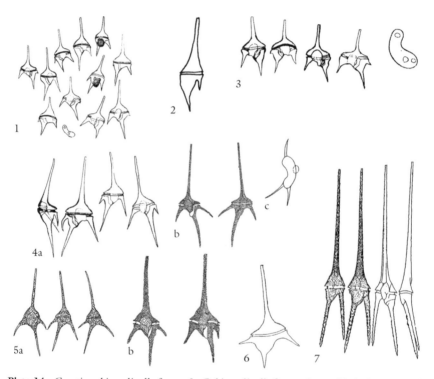

Plate 14. *Ceratium hirundinella* forms. 1. *C. hirundinella* f. *austriacum* (Zederbauer 1904b, Taf 5, Figs. 13–25). 2. *C. hirundinella* f. *brachyceroides* with two posterior horns, one incurved (redrawn from Schröder 1918, Fig. 2). 3. *C. hirundinella* f. *carinthiaceum* (Zederbauer 1904a, Taf 5, Figs. 1, 2, 4, 6, 7). 4. *C. hirundinella* f. *piburgense,* postcingular horns straight, all relatively long, outwardly splayed. (a) (Zederbauer 1904b, Taf 5, Figs. 8, 9, 11, 12) (b) (Eddy 1930, Pl. 35, Figs. 58a,b) (c) Bottom view with horns splayed. 5. *C. hirundinella* f. *robustum.* (a) (Amberg 1903, 83) (b) (Eddy 1930, Pl. 35, Figs. 58c,d). 6. *C. hirundinella* f. *scotticum,* three posterior horns about equal in length, short (Bachmann 1911, Fig. 54). 7. *C. hirundinella* f. *yuannense,* long narrow cell with two posterior horns (Skuja 1937, Taf 3, Figs. 9–12).

2b. Cell shorter, antapical horn incurved .... *C. h.* form *brachyceroides*
3a. Postcingular horns straight 4
3b. Postcingular horns curved .... *C. h.* form *tridenta*
4a. Posterior horns about equal in length 5
4b. Posterior horns short-long-medium 6
5a. Horns short, body squat .... *C. h.* form *scotticum*
5b. Horns long, straight, widely splayed .... *C. h.* form *piburgense*
6a. Horns splayed 7
6b. Horns directed downward 8
7a. Central, antapical horn distinctly longer .... *C. h.* form *hirundinella*
7b. Antapical horn somewhat longer than medium .... *C. h.* form *austriacum*
8a. One postcingular horn reduced, wide body .... *C. h.* form *carinthiacum*
8b. Apical horn at an angle to body .... *C. h.* form *robustum*

Ceratium hirundinella f. *hirundinella*

Three posterior horns of different lengths, directed straight down to slightly splayed. **Figs. 45, 46; CP5(2)**

EXTERNAL FEATURES: Two straight, outward-pointing postcingular horns, the right well developed, the left with varying degrees of development from (none?) small to almost equaling the right. The body is approximately equally divided into epitheca and hypotheca (hypotheca slightly larger). Cysts have four horns.

SIZE: Total length 188–247 μm.

Ceratium hirundinella f. *austriacum*

(Zederbauer) Bachman 1911, 73, Fig. 51.

BASIONYM: *Ceratium austriacum* Zederbauer 1904b, 168, Taf. 5, Figs. 13–25.

Two outward spreading postcingular horns, the left may be rudimentary. It is similar to f. *scotticum* but its horns are less splayed, and it is more balanced in epitheca and hypotheca than f. *scotticum*. It looks like a stocky f. *hirundinella* with a shorter antapical horn. **PL14(1)**

SIZE: 115–145 μm long by 45–60 μm wide.

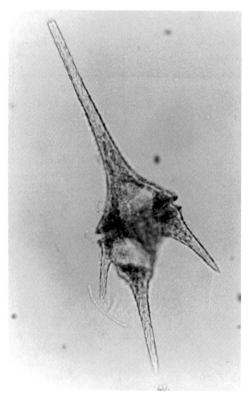

Figures 45, 46. *Ceratium hirundinella.* Note straight apical horn arising from shoulders; three straight, downward-pointing posterior horns, the central one longest.

Ceratium hirundinella f. *brachyceroides*

Schröder 1918, 226, Fig. 2.

Distinguished from the other forms by its inward curving antapical horn. The single postcingular horn may also be slightly incurved (Schröder 1918). **PL14(2); Figs. 47, 48**

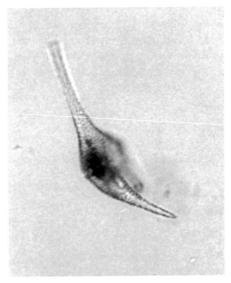

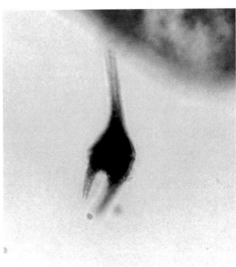

Figures 47, 48. *Ceratium hirundinella* f. *brachyceroides.* Note incurved hypothecal horn.

SIZE: Width up to 45 μm (Schröder 1918).

LOCATION REFERENCE: OH (SC 970716-12), Michoacán (Tafall 1941).

Ceratium hirundinella f. *carinthiacum*

(Zederbauer) Bachmann 1911, 73, Fig. 50.

BASIONYM: *Ceratium carinthiacum* Zederbauer 1904a, 127–128, Taf. 5, Figs. 1–7.

Short and wide with usually one short, slightly divergent postcingular horn. It is similar to form *austriacum* but the postcingular horns of the former are more divergent; the epitheca is reminiscent of f. *piburgense*. **PL14(3)**

SIZE: 100–150 μm long by 50–60 μm wide (Zederbauer 1904a).

Ceratium hirundinella f. *piburgense*

(Zederbauer) Bachmann 1911, 73, Fig. 52.

BASIONYM: *Ceratium piburgense* Zederbauer 1904b, 167–168, Taf. 5, Figs. 8–12.

All three of its posterior horns are long and widely splayed; it has the most compact body (most obvious shoulders). Cysts have four extended horns (Wall and Evitt 1975, Pl. 1, Figs. 6, 7). The antapical horn tends to look like a peg leg, wide near the body, then abruptly tapering to a narrower section. **PL14(4); Figs. 49, 50; CP5(4)**

SIZE: Total length 225–322 μm.

Ceratium hirundinella f. *robustum*

Amberg 1903, 83.

Form *robustum*, like f. *austriacum*, is similar to the type form. It has two divergent postcingular horns, and is described as bent in side view (Schiller 1937). **PL14(5)**

SIZE: 180–200 μm long by 65–75 μm wide (Amberg 1903).

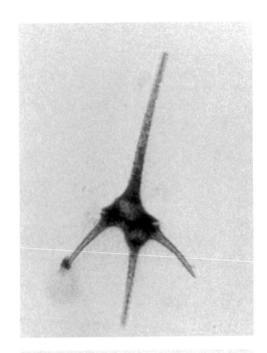

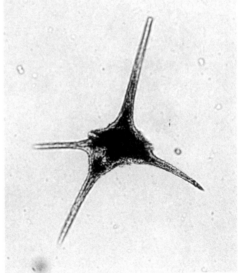

Figures 49, 50. *Ceratium hirundinella* f. *piburgense.* Note straight apical horn arising abruptly from shoulders; three long, straight, outward-pointing posterior horns.

Ceratium hirundinella f. *scotticum*

Bachmann 1911, 75, Fig. 54.

Two short, widely spreading postcingular horns. It is similar to f. *austriacum* but its postcingular horns are more spreading and a larger proportion of the cell is epitheca; length 210 µm (Bachmann 1911). **PL14(6); Fig. 51**

SIZE: 183 µm long.

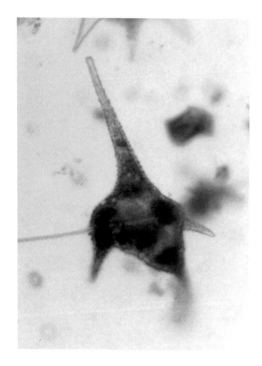

Figure 51. *Ceratium hirundinella* f. *scotticum*. Three posterior horns, the two postcingular horns short.

Ceratium hirundinella f. *yuennanense*

(Skuja) Huber-Pestalozzi 1950.

BASIONYM: *Ceratium handelii* Skuja 1937, 48, Taf. 3, Figs. 9–12.

The longest of the forms (375–443 µm) with two approximately equal downward-pointing posterior horns. Not reported from North America. **PL14(7)**

Ceratium hirundinella f. *tridenta* n. forma

Postcingular horns downcurved, body wide, short. **Figs. 52–56; CP5(5,6)**

EXTERNAL FEATURES: Body wide and short, long straight apical horn, antapical horn straight, may be at an obtuse angle with apical horn. Postcingular horns similar in length and downcurved. No other form has curved postcingular horns. Name based on the trident appearance of cell.

SIZE: Total length (apical horn to antapical horn) 283–309 µm, maximum width (between postcingular horns) 100–114 µm, body width 43–56 µm, body length 20–38 µm.

LOCATION REFERENCE: Lake LaSuAn, Williams County, Ohio, 6/27/97; a northern Wisconsin lake (Gina LaLiberte pers. comm.); and NY (John Hall, pers. comm.).

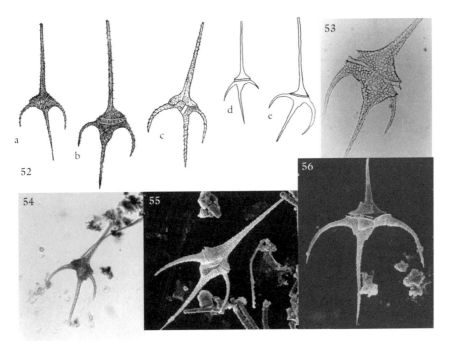

Figure 52. *Ceratium hirundinella* f. *tridenta* n. forma. Postcingular horns downcurved. (a) Ventral (b) Dorsal (c) Ventral (d, e) Dorsal (drawn from micrographs) (a, b courtesy of Mary Ann Rood).

Figures 53, 54. *Ceratium hirundinella* f. *tridenta*. Curved, downward-pointing postcingular horns (LM).

Figure 55. Ventral (SEM).

Figure 56. Possible release of cyst or regrowth of apical half.

Ceratium rhomvoides

Hickel 1988, 49–54, Figs. 4–6.

SYNONYM: *Ceratium hirundinella* f. *gracile* Bachmann 1911, 73, Fig. 53.

Straight apical horn, two downward-pointing posterior horns. **PL13(6); Figs. 57–59; CP5(7–9)**

EXTERNAL FEATURES: Cell with straight apical horn arising from shouldered body, with apical pore and pore plate, two downward-pointing posterior horns. Fourth apical plate does not reach apex. Cingulum without displacement, sulcus strongly concave, Sa and Sd similar size, Sp narrow and parallel to 2''''. Asexual reproduction by division of parental cell along predetermined sutures; sexual reproduction by incorporation of male gamete into female. The epitheca is approximately as long as the hypotheca, and the dorsal epitheca is described as not convex (Popovský and Pfiester 1990). The cyst is roundish with three short horns (Wall and Evitt 1975, Pl. 1, Fig. 9). In reviewing possible synonymies Hickel (1988) noted that Entz (1925) had illustrated *Ceratium hirundinella* f. *gracile* with a 4' not reaching the apex. I have reviewed micrographs of cells I had called *Ceratium hirundinella* f. *gracile* and found the 4' not reaching the apex.

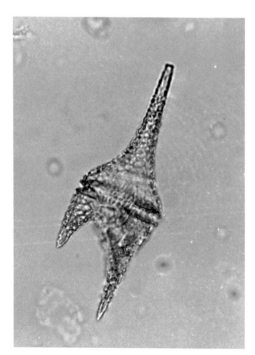

Figure 57. *Ceratium rhomvoides* with two straight, downward-pointing posterior horns; dorsal empty cell (LM).

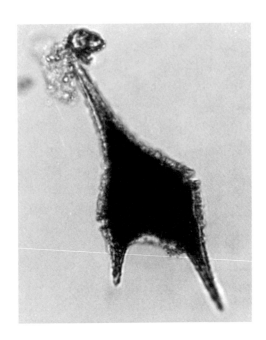

Figure 58. Cell stained with iodine (LM).

Figure 59. Dorsal (SEM).

INTERNAL FEATURES: Photosynthetic with brown plastids.

SIZE: 123–203 μm long by 32–56 μm wide (Hickel 1988).

ECOLOGICAL NOTES: Found in the summer with *C. hirundinella* and *C. furcoides* (Hickel 1988).

Dinosphaera

Kofoid and Michener 1912, 23.

TYPE SPECIES: *Dinosphaera palustris* (Lemmermann) Kofoid and Michener 1912, 23–28.

Plate formula: no apical pore, 3', 1a, 6'', C6, T, S4, 5''', 1''''. Freshwater genus.

Dinosphaera palustris

(Lemmermann) Kofoid and Michener 1912, 23–28.

BASIONYM: *Gonyaulax palustris* Lemmermann 1907, 296, Figs. 1–5.

SYNONYM: *Glenodinium palustre* (Lemmermann) Schiller 1937, 99.

Round cell; three large plates (2', 3', 1a) at apex; one antapical plate. **Fig. 60**

EXTERNAL FEATURES: Cell almost spherical, smooth, cingulum median, displaced one cingulum width, sulcus with slight penetration into epitheca, not reaching antapex, single antapical plate and large 5''' plate reaching antapex (Kofoid and Michener 1912).

INTERNAL FEATURES: Chromatophores small, many, parietal (Tiffany and Britton 1952); no eyespot (Forest 1954a); yellow oil drops present (Kofoid and Michener 1912).

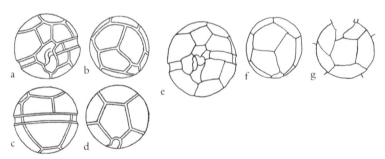

Figure 60. *Dinosphaera palustris.* (a) Ventral (b) Apical (c) Dorsal (d) Antapical views (Lemmermann 1907) (e) Ventral (f) Apical (g) Antapical (modified from Kofoid and Michener 1912, Figs. 6, 7, 8).

SIZE: 27–34 μm long, diameter 25–31 μm (Kofoid and Michener 1912).

ECOLOGICAL NOTES: Reported from softwater lakes and acid bogs (Prescott 1951b), blooming in a pool (Prescott and Croasdale 1937), in the euplankton of a tundra pond (Sheath and Hellebust 1978).

Durinskia

Carty and Cox 1986, 200.

TYPE SPECIES: *Durinskia baltica* (Levander) Carty and Cox 1986, 200.

Plate pattern: apical pore, pp, cp, 4′, 2a, 6″, 5C (T + 4C), 4S, 5‴, 2⁗

Plate pattern necessary for identification; golden, plates thin, without ornamentation, eyespot present in sulcus. Freshwater.

Durinskia dybowski

(Wołoszyńska) Carty comb. nov.

BASIONYM: *Peridinium dybowski* Wołoszyńska 1916, Taf. 13, Fig. 9–14.

SYNONYMS: *Durinskia baltica* (Levander) Carty and Cox 1986, 200; *Glenodinium balticum* Levander 1894.

Upper cingular margins strongly defined; cell round to oval; brownish; eyespot present; small (25 μm), thinly thecate. **Figs 61–63; CP6(1,2)**

EXTERNAL FEATURES: Plate pattern: apical pore, pp, cp, 4′, 2a, 6″, 5C (T + 4C), 4S, 5‴, 2⁗; round to oval cell, dorsoventral compression noticeable when cells swimming; plates thin, smooth, sutures difficult to see, 4′ plate large, 1a small, 2a large and across the back of the cell; cingulum median with little displacement, slightly indented with distinct edges, upper cingular margin strongly defined, may extend further than lower; sulcus only in the hypotheca and not extending to antapex, Sd plate forms a flap over sulcal area; antapical plates about equal in size.

INTERNAL FEATURES: Plastids numerous, golden to brown; nucleus central; may have red accumulation bodies; eyespot in sulcus.

SIZE: 25–36 μm long by 21–32 μm wide, 25–32 μm thick.

ECOLOGICAL NOTES: I have found this species in the summer in a mixed population in a small pond (TN), and almost unialgal in a small pond (OH). A report from Mexico City finds heavy growth year-round in shallow, eutrophic channels with cysts present in sediments in March (Rosaluz Tavera, pers. comm.)

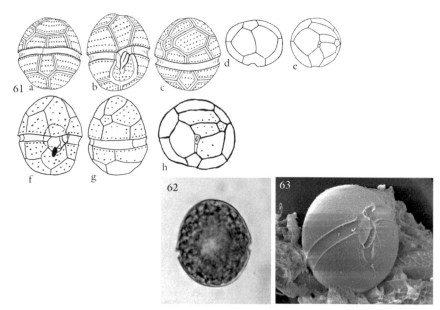

Figure 61. *Durinskia dybowski.* (a) Dorsal (b) Ventral (c) Left dorsal (d) Antapical (e) Apical (Wołoszyńska 1916, Pl. 13, Figs. 9–13) (f) Ventral with eyespot and central nucleus (g) Dorsal (h) Apical (Carty 2003, Fig. 3Fa,b,c).

Figure 62. *Durinskia dybowski* cell; note central nucleus (LM).

Figure 63. Ventral view (SEM) (Carty and Cox 1986).

Entzia

Lebour 1922, 808.

TYPE SPECIES: *Entzia acuta* (Apstein) Lebour 1922, 808.

Plate formula: apical pore, pp, 4', 2a, 7", C3, 5''', 1''''; 3' plate small, intercalary plates large.

Cell wider than long, tapering acutely from the cingulum to a distinct apical region and pore. Hypotheca rounded. Freshwater.

Entzia acuta

(Apstein) Lebour 1922, 808.

BASIONYM: *Glenodinium acutum* Apstein 1896, 152.

SYNONYMS: *Diplopsalis acuta* (Apstein) Entz 1904.

Prominent sulcal list extending past the antapex, nonphotosynthetic. **PL15(1); Figs. 64, 65**

EXTERNAL FEATURES: Cell squat, epitheca tapering

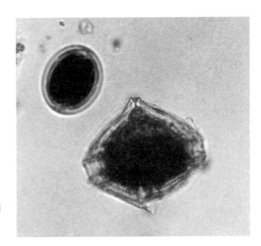

Figure 64. *Entzia acuta.* Note apical pore, distinctive cingular list; cell stained with iodine (LM).

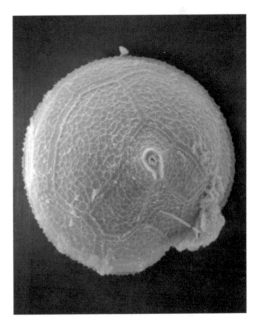

Figure 65. Apical view; note apical pore, four small apical plates, two large apical intercalary plates, almost round circumference (SEM).

to extended apex; hypotheca flattened, sulcus not extending into the epitheca, deep, with parallel sides extending to the antapex. Deep cingular groove enhanced by flange on cingular margins of pre- and postcingular plates, prominent sulcal list on the left side extending past the antapex, light reticulate ornamentation. Bourrelly (1970) illustrated both six and seven precingular plates.

INTERNAL FEATURES: Nonphotosynthetic cell, may have grainy appearance to cytosol.

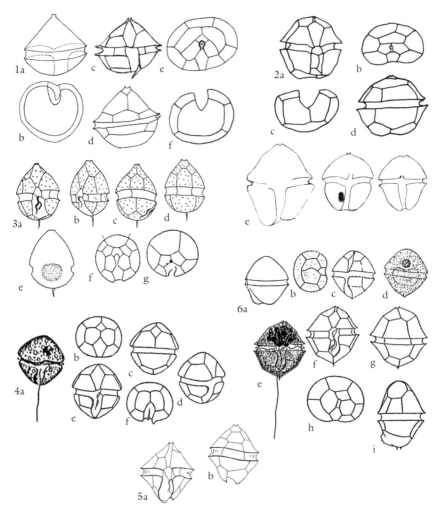

Plate 15. Thecate, nonphotosynthetic species. 1. *Entzia acuta.* (a) Dorsal view of empty cell showing cingulum and sulcus (b) Hypotheca with single antapical plate (Apstein 1896, Fig. 54 as *Glenodinium acutum*) (c) Ventral view; note extended sulcal list (d) Dorsal (e) Epitheca (f) Hypotheca (orig.). 2. *Glochidinium penardiforme.* (a) Ventral (b) Epitheca (c) Hypotheca (d) Dorsal; note bilobed posterior (redrawn and modified from Boltovskoy 1999, Figs. 3–6, with permission) (e) Three proportional sketches (Victor W. Fazio, with permission). 3. *Parvodinium goslaviense.* (a) Ventral; note single antapical spine (b) Left ventral (c) Left side (d) Dorsal (e) Nucleus in hypotheca (f) Apical (g) Antapical (Wołoszyńska 1916, Pl. 10, Figs. 18–24). 4. *Peridiniopsis edax.* (a) Ventral (b) Epitheca with plates (c) Dorsal (d) Right side (e) Ventral (f) Hypotheca (Thompson 1950, Figs. 52–57). 5. *Peridinium achromaticum.* (a) Ventral (b) Dorsal (Wołoszyńska 1928, Pl. 12, Figs. 5, 6). 6. *Tyranodinium berolinense.* (a) Left side (b) Hypotheca (c) Left sulcal (d) Dorsal view with central nucleus (Lemmermann 1910, 658, Figs. 17–20) (e) Ventral view of cell, sulcus not to antapex, antapical spine (f) Ventral with plates (g) Dorsal (h) Apical (i) Left lateral view (Thompson 1950, Figs. 29–32, 34).

SIZE: Cells 30–49 μm long by 39–44 μm wide.

ECOLOGICAL NOTES: Collected from the plankton of a large lake (September, OK, Pfiester and Terry 1978).

Glenodiniopsis

Wołoszyńska 1916, 278.

TYPE SPECIES: *Glenodiniopsis steinii* Wołoszyńska 1916, 278–279, Taf. 11, Figs. 30–36.

No apical pore, plate formula 2′, 4a, 8″, 7‴, 2⁗.

Cell ovoid, epitheca larger than hypotheca; plates thin and difficult to see, irregularly polygonal; photosynthetic. One species known from North America, *G. steinii*.

Glenodiniopsis steinii

Wołoszyńska 1916, 278–279, Taf. 11, Figs. 30–36.

SYNONYM: *Glenodinium uliginosum* Schilling 1913.

Cingulum much more pronounced on one side, distinctly offset; golden. **Fig. 66; CP6(3)**

EXTERNAL FEATURES: Oval-shaped cell, some dorsoventral compression; cingular groove deeper on left side, cingulum displaced two cingulum widths; plates smooth, slightly punctate at sutures; two large apical plates, cover half of epitheca.

INTERNAL FEATURES: Nucleus central; chloroplast single, lobed, appearing as numerous discoid chloroplasts; no eyespot (Highfill and Pfiester 1992a).

SIZE: 26–50 μm long by 26–35 μm wide.

ECOLOGICAL NOTES: Collected from pond (MN, Highfill and Pfiester 1992a), from sediments in a small pool (NY, John Hall, pers. comm., with a bloom of *Hyalotheca dissiliens*).

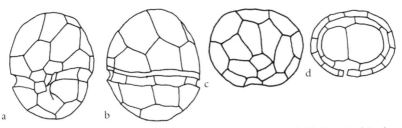

Figure 66. *Glenodiniopsis steinii.* (a) Ventral (b) Dorsal (c) Apical (d) Antapical (redrawn from Highfill and Pfiester 1992b, Fig. 7).

Glochidinium

Boltovskoy 1999, 98–99.

TYPE SPECIES: *Glochidinium penardiforme* (Lindemann) Boltovskoy 1999.

Genus erected based on *Peridinium penardiforme*. Plate formula: apical pore, pp, cp, 3', 1a, (or 4'-0a), 6", C3, S4, 5''', 2''''. Three cingular plates distinguish it from related genera as does a small Sp plate.

One species reported from North America, *Glochidinium penardiforme* (see Boltovskoy 1999 for additional species from Argentina).

Glochidinium penardiforme

(Lindemann) Boltovskoy 1999, 98–99.

BASIONYM: *Peridinium penardiforme* Lindemann 1918, 126, Figs. 10–15.

SYNONYM: *Glenodinium penardiforme* (Lindemann) Schiller 1937.

Colorless, thecate cell with deeply bilobed posterior. **PL15(2); Fig. 67; CP6(7)**

EXTERNAL FEATURES: Plate formula pp, cp, 3', 1a, (or 4'-0a), 6", C3, S4, 5''', 2''''; epitheca tapers from cingulum to apical pore; apical pore surrounded by pore plate with ridge around pore, canal plate; plates surrounding pore (2', 3', 4') with raised ridge around pore and canal plates, gives a keyhole shape; hypotheca with flattened antapical region, strong dorsoventral compression; plates not obvious; deeply incised cingulum without displacement. Sulcus penetrates slightly into epitheca and there is a deep groove to the antapex. Thompson illustrates a symmetrical arrangement of epithecal plates including a large 3' plate and smaller 2' and 4' (Thompson 1950, KS). Cells collected from Nebraska have

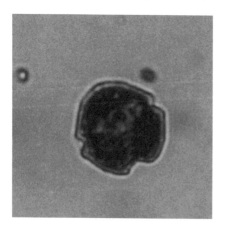

Figure 67. *Glochidinium penardiforme*. Cell stained with iodine; note strongly bilobed posterior (LM).

large 2' and 4' plates and a smaller 3' plate. Antapical plates equal in size.

INTERNAL FEATURES: Without plastids (or with); nucleus round, central; may have grainy appearance to cytosol.

SIZE: 17–25 μm long by 17–23 μm wide, 9–11 μm thick.

Gonyaulax

Diesing 1866.

TYPE SPECIES: *Gonyaulax spinifera* (Claparède and Lachmann) Diesing 1866.

Plate pattern: apical pore, 4', 5", 6C, 3S, 6''', 2''''. Deep sulcal penetration into epitheca, not reaching antapex.

Most species marine, also an extensive fossil history.

Key to freshwater species of *Gonyaulax*

1a: Sulcus vertical... *G. apiculata*
1b: Sulcus angled ....*G. spinifera*

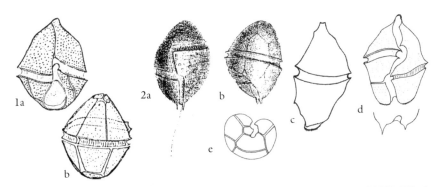

Plate 16. Species of *Gonyaulax*. 1. *G. apiculata*. (a) Ventral (b) Dorsal (Penard 1891, Tab. 3, Figs. 4, 7). 2. *G. spinifera*. (a) Ventral (b) Side (Claparède and Lachman 1859, Pl. 20, Figs. 4, 5) (c) Side (d) Ventral (e) Hypotheca (Carty 2003, Figs. 3Ba,c).

Gonyaulax apiculata

(Penard) Entz 1904.

BASIONYM: *Peridinium apiculatum* Penard 1891, 51, Pl. 3, Figs. 3–13.

Cingulum offset but sulcus vertical and almost reaching apex. **PL16(1)**

EXTERNAL FEATURES: Ovate cell; epitheca apiculate; hypotheca rounded; cingulum offset one cingulum width, the ventral left portion descending; sulcus deeply indented, vertical, slightly penetrating epitheca and continuing as a slit/suture to the right of the apex, widening in the hypotheca to antapex; punctate ornamentation. Cell ecdyses along line formed by slit/suture and left sulcal margin.

INTERNAL FEATURES: Nucleus central, golden-brown chloroplasts.

SIZE: 30–62 μm by 29–57 μm (Popovský and Pfiester 1990).

LOCATION REFERENCE: BC (Stein and Borden 1979 as *Gonyaulax* sp.), NC (Whitford and Schumacher 1969).

Gonyaulax spinifera

(Claparède et Lachmann) Diesing 1866

BASIONYM: *Peridinium spiniferum* Claparède et Lachmann 1858, 1859, 52, Pl. 20, Figs. 4, 5.

Twisted sulcus and protracted apex. **PL16(2); Figs. 68, 69**

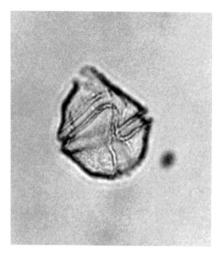

Figure 68. *Gonyaulax spinifera.* Empty cell; note twisted sulcus.

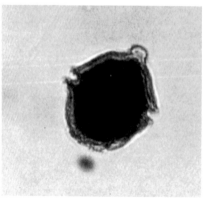

Figure 69. Stained cell; note distinctive apical prominence.

EXTERNAL FEATURES: Cingulum offset two cingulum widths, twisted appearance to sulcus due to overlapping ends of cingulum. Thick plates with punctate ornamentation, antapex with spines/list.

INTERNAL FEATURES: Golden cells.

SIZE: 28 μm long.

ECOLOGICAL NOTES: Collected from a pond on eastern Long Island (NY, V. Fazio, pers. comm.).

LOCATION REFERENCE: NY (VF Sagaponack 8-16-97).

Hemidinium

Stein 1883.

TYPE SPECIES: *Hemidinium nastum* Stein 1878.

Incomplete cingulum, golden plastids.

Oval, golden, freshwater motile cells with incomplete cingulum giving the cell a slashed appearance. Forms a round resting cell within (layered) mucilage.

There has long been a controversy about the relationship between *Hemidinium* and *Gloeodinium* since *Hemidinium* produces *Gloeodinium*-like resting cells and *Gloeodinium* produces *Hemidinium*-like motile cells (see "Taxonomy" in the introduction). Cell division has been reported to occur while cells are motile and in an approximate transverse plane (Smith 1933, Bicudo and Skvortzov 1970).

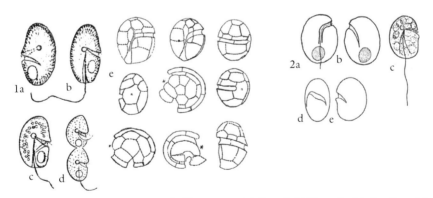

Plate 17. Species of *Hemidinium*. 1. *H. nasutum*. (a) Dorsal (b) Ventral with flagellum (c) Ventral (d) Stage seen only twice by Stein (1883, Taf 2, Figs. 23–26) (e) Plate pattern (Wołoszyńska 1925b, Fig.4). 2. *H. ochraceum*. (a) Ventral; note position of cingulum above median (b) Dorsal (Levander 1900a, Fig. 2) (c) Ventral; note slightly swollen lower lip of cingulum (d) Side view of entire cingulum (e) Dorsal view (Thompson 1947, Plate 2, Figs. 4–6).

Key to species of *Hemidinium*

1a. Cingulum median .... *H. nasutum*
1b. Cingulum above median .... *H. ochraceum*

Hemidinium nasutum

Stein 1883, Taf. 2, Figs. 23–26.

Dorsoventrally flattened cell with median incomplete cingulum. **PL17(1); Figs. 70, 71; CP6(4,5)**

EXTERNAL FEATURES: Elliptical cell with rounded poles, dorsoventrally compressed; thin theca with plates not visible, though plate pattern has been determined to be 6′, 6″, 5‴, 1p, 1⁗ (Wołoszyńska 1925b, Pfiester and Highfill 1993), cingulum median, slashlike, descending to the left from a midventral position, ending middorsally, sulcus only in the hypotheca, reaching antapex.

INTERNAL FEATURES: No eyespot; golden plastids fusiform-elongate; nucleus posterior.

SIZE: 23–32 μm long by 11–22 μm wide, 8–11 μm thick.

ECOLOGICAL NOTES: To give an idea of where and when *Hemidinium nasutum* has been collected, the following are some of the reports: MD, summer (July–September), creek, pond, swamp (Thompson 1947); NC, November, pond (Whitford and Schumacher 1969); WI, common in the nearshore plankton of lakes (Prescott 1951b); MN, dystrophic ponds and planktonic

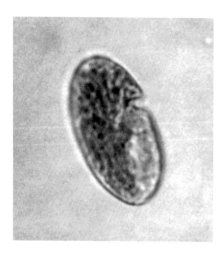

Figure 70. *Hemidinium nasutum* note slashed appearance of incomplete cingulum (*arrow*), ventral view (Carty 2003, Fig. 8K).

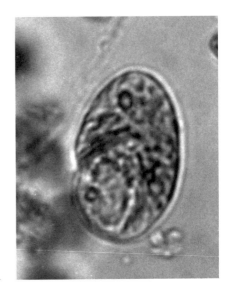

Figure 71. Dorsal view.

in a small lake (Meyer and Brook 1969); IA, March, quarry pond (Prescott 1927); TN, swamp (Forest 1954a); OK, July, weedy pond (Pfiester and Terry 1978); TX, year-round, pond (Carty 1986).

Hemidinium ochraceum

Levander 1900a, 103, Fig. 2; 104 as *Hemidinium ochraceus.*

Incomplete cingulum above median with lower margin slightly inflated. **PL17(2)**

EXTERNAL FEATURES: Cell oval, broadly rounded at apex and antapex; cingulum incomplete, begins above median on ventral face, slants downward, continuing about a third of the way across the dorsal face, lower margin with distinctive rounded lip; sulcus not reaching antapex (Thompson 1947).

INTERNAL FEATURES: 5–11 large, almost circular, golden to greenish plastids; the anterior part of the cell is red brown, the posterior green (Thompson 1947); nucleus in hypotheca.

SIZE: 28–30 μm long by 21 μm wide, 15–18 μm thick.

ECOLOGICAL NOTES: Collected from a quarry pool in December (Thompson 1947).

LOCATION REFERENCE: MD (Thompson 1947).

Jadwigia

Moestrup, Lindberg, and Daugbjerg in Lindberg et al. 2005, 432.

TYPE SPECIES: *Jadwigia applanata* Moestrup, Lindberg, and Daugbjerg in Lindberg et al. 2005, 432.

Roundish photosynthetic cell with eyespot, with numerous, thin, hexagonal plates.

This genus was extracted from *Woloszynskia* based on *W. neglecta* (as illustrated by Wołoszyńska 1917 and Lindemann 1929), for species with numerous, thin, usually hexagonal plates; an apical line of narrow plates (ALP); an eyespot of red globules not associated with a chloroplast; and a round cyst (Lindberg et al. 2005).

Jadwigia applanata

Moestrup, Lindberg, and Daugbjerg in Lindberg et al. 2005, 432.

SYNONYMS: *Woloszynskia neglecta* (Schilling) Thompson 1950, 290; *Gymnodinium neglectum* (Schilling) Lindemann 1929, 60-61; *Glenodinium neglectum* Schilling 1891a; *Glenodinium neglectum* Schilling as illustrated by Wołoszyńska 1917, Pl. 11 Fig. 5, Pl. 12, Figs. 10–11; *Woloszynskia limnetica* Bursa 1958, 299–301, Figs. 1–5.

Numerous small hexagonal plates, nucleus in epitheca, round cyst. **PL12(2)**

EXTERNAL FEATURES: Ovate to spherical; slightly dorsoventrally flattened; hypotheca hemispherical; epitheca slightly conical; cingulum median with displacement; sulcus not entering epitheca and only slightly entering hypotheca; theca of numerous small hexagonal plates; SEM micrographs reveal a single narrow line of rectangular cells beginning on the ventral face of the epitheca and extending over the apex; cysts spherical, smooth, with eyespot (may be double as result of fused gametes), 24.5–27.7 μm diameter (Lindberg et al. 2005).

INTERNAL FEATURES: Elongate eyespot in sulcus; nucleus sausage-shaped in right side of epitheca; plastids numerous, peripheral, golden to brown.

SIZE: 32–35 μm long by 25–30 μm wide (Forest 1954a).

Kansodinium

Carty and Cox 1986, 197.

TYPE SPECIES: *Kansodinium ambiguum* (Thompson) Carty and Cox 1986, 197.

CP1. Ecological. 1. Dinoflagellate bloom, Texas, summer 2009. 2. Dinoflagellate bloom, Texas, summer 2009 (1 and 2 courtesy of Cheryl Gilpin, MS, Oceanography, Texas A&M University, seacheyl@gmail.com). 3. Field conditions, looking at algae immediately after collection, Great Smoky Mountains National Park (Susan Carty, John D. Hall) 4. Beaver pond, Maine. 5. Muskeg area, Juneau, Alaska. 6. Water lily pond.

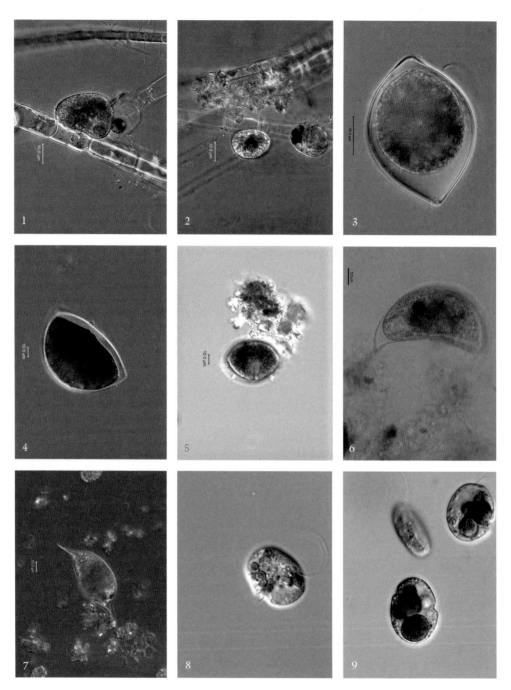

CP2. Athecate species. 1. *Cystodinedria inermis* on filamentous green alga, shortly after ingesting cell contents (note empty cell beneath). 2. *Cystodinedria inermis* on filamentous green alga, orange phase (note empty cell beneath). 3. *Cystodinium bataviense*. Note lemon-shaped envelope, golden cell within. 4. *Cystodinium bataviense*. Note indented cingulum on cell within. 5. *Cystodinium bataviense*. Note golden cell with red accumulation bodies. 6. *Cystodinium iners*. Note curved cell and pointed ends (courtesy of John D. Hall). 7. *Cystodinium iners*. Cell more *Gymnodinium*-like within envelope, eyespot visible (courtesy of John D. Hall). 8. *Esoptrodinium gemma* with plastids, no eyespot (courtesy of Matthew W. Parrow). 9. *Esoptrodinium gemma* with red accumulation bodies, *Rhodomonas* cell (prey) present (courtesy of Matthew W. Parrow).

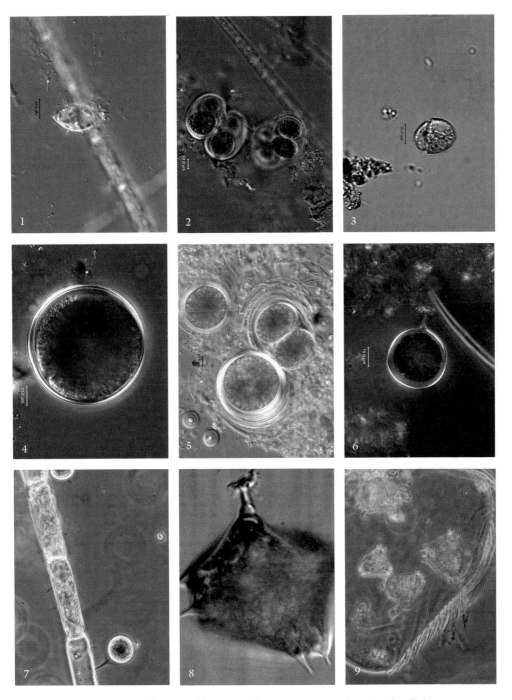

CP3. Athecate species. 1. *Dinococcus bicornis* on filamentous green alga. 2. *Gloeodinium montanum*. Golden cells in sheaths. 3. *Opisthoaulax vorticella* with eyespot in sulcus, slight indentation of sulcus into epicone, unequal-sized ventral hypocone portions; this species is heterotrophic, so pigment may be from ingested prey. 4. *Hypnodinium?* Round cell in sheath, with accumulation bodies. 5. *Rufusiella insignis.* Note concentric layers of envelope. 6. *Stylodinium globosum.* Detached cell with stalk and holdfast. 7. *Stylodinium globosum* attached to filamentous alga (cell beneath is empty). 8. *Tetradinium javanicum* cell with plastids, attachment stalk, and paired spines at corners. 9. *Tetradinium javanicum.* A few cells on empty zooplankter shell.

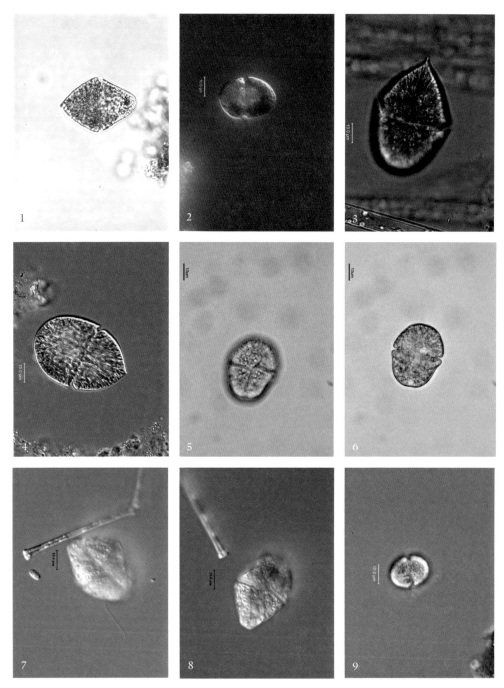

CP4. *Gymnodinium* species. 1. *Gymnodinium acidotum* with acute posterior. 2. *Gymnodinium aeruginosum* with rounded posterior; note nucleus in hypocone. 3. *Gymnodinium caudatum*, similar to *G. fuscum* but with extended tail. 4. *Gymnodinium fuscum*. Broadly rounded epicone, tapered hypocone, large cell. 5. *Gymnodinium inversum*. Hypocone slightly larger, extension of sulcus into epicone 6. *Gymnodinium inversum*. No eyespot. 7. *Gymnodinium thompsonii*. Strongly cruciate cingulum and sulcus. 8. *Gymnodinium thompsonii* 9. *Gymnodinium aeruginosum*. Focus on sulcus extending to antapex.

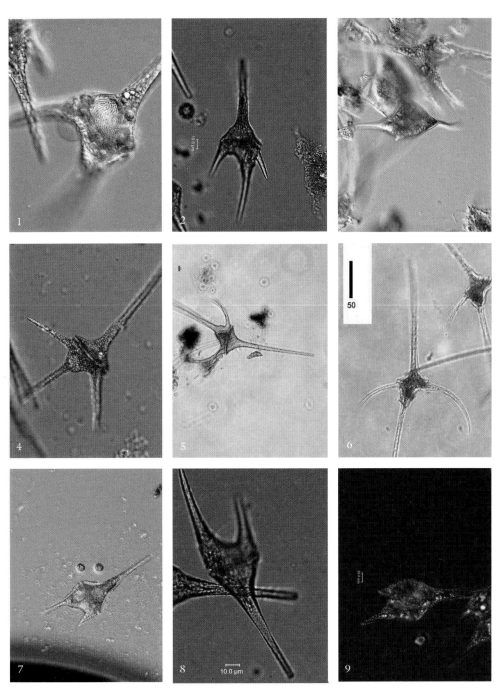

CP5. *Ceratium.* 1. *Ceratium hirundinella.* Note large nucleus with permanently condensed chromosomes, golden chloroplasts. 2. *Ceratium hirundinella* f. *hirundinella.* Three downward-pointing posterior horns, the middle the longest. 3. *Ceratium hirundinella* cyst. 4. *Ceratium hirundinella* f. *piburgense.* Note straight, long, outward-pointing posterior horns, peg-leg antapical horn. 5. *Ceratium hirundinella* f. *tridenta.* Note curved postcingular horns (courtesy of John D. Hall). 6. *Ceratium hirundinella* f. *tridenta.* Note curved postcingular horns (courtesy of Gina LaLiberte). 7. *Ceratium rhomvoides.* Two downward-pointing posterior horns. 8. *Ceratium rhomvoides.* Two downward-pointing posterior horns. 9. *Ceratium rhomvoides.* Note elongate sulcal opening.

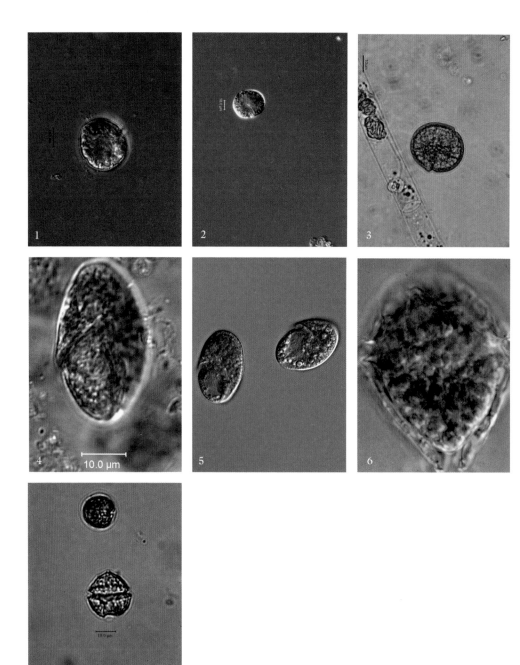

CP6. Thecate taxa D–L. 1. *Durinskia dybowski*. Red accumulation body at apex, small eyespot in sulcus. 2. *Durinskia dybowski*, showing details of sulcus. 3. *Glenodiniopsis steinii* (courtesy of John D. Hall). 4. *Hemidinium nasutum*. Golden cell with cingulum incomplete and looking like a slash. 5. *Hemidinium nasutum*. Two cells (courtesy of Matthew W. Parrow). 6. *Lophodinium polylophum*. Note eyespot, chloroplasts, and opening at apex. 7. *Glochidinium penardiforme*. Colorless, with bilobed posterior.

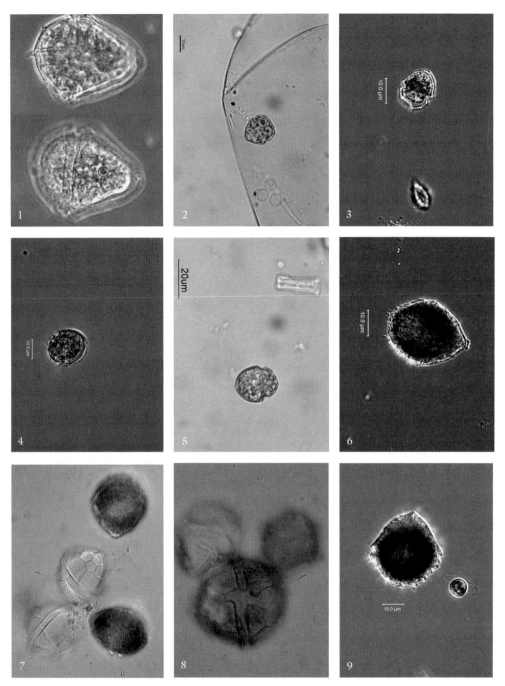

CP7. 1. *Naiadinium biscutelliforme.* Single large antapical spine. 2. *Parvodinium inconspicuum.* Pentagonal cell (courtesy of John D. Hall). 3. *Parvodinium inconspicuum.* This hypotheca with small spines. 4. *Parvodinium umbonatum.* Rounded epitheca. 5. *Parvodinium umbonatum.* Epitheca larger than hypotheca (courtesy of John D. Hall). 6. *Peridiniopsis quadridens.* Antapical spines and chimney-like apical pore. 7. *Thompsodinium intermedium.* Golden chloroplasts, empty cells with light striations on plates. 8. *Thompsodinium intermedium.* Ventral view. 9. *Woloszynskia reticulata.* Note bumpy margin of hypotheca due to heavy plate margins. (1, 7, and 8 courtesy of Cheryl Gilpin, MS, Oceanography, Texas A&M University, seacheyl@gmail.com.)

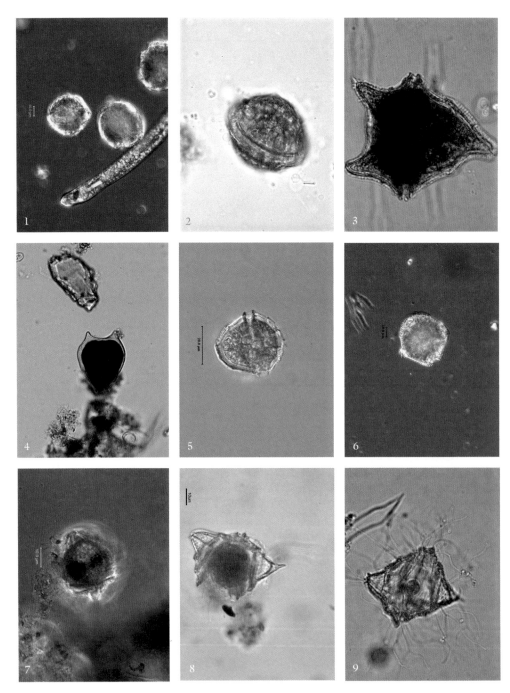

CP8. 1. *Peridinium gatunense* with a *Euglena* for color comparison. 2. *Peridinium gatunense.* Saturn-like appearance of cell. 3. *Peridinium limbatum.* Cell stained with iodine. 4. *Peridinium limbatum* cyst. 5. *Peridinium volzii.* 6. *Peridinium willei* with plate margin extensions (lists) at apex and antapex. 7. *Peridinium wisconsinense.* 8. *Peridinium wisconsinense.* 9. *Peridinium wisconsinense.*

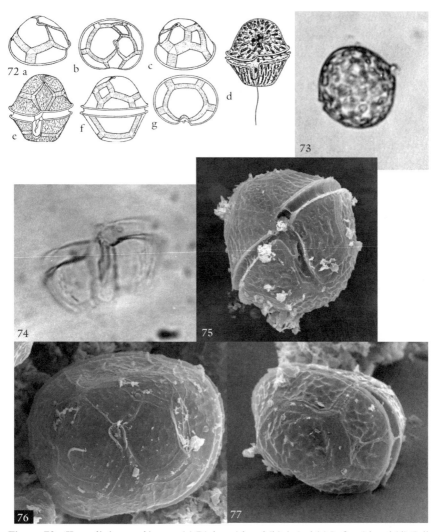

Figure 72. *Kansodinium ambiguum.* (a) Right epithecal (b) Apical (c) Left epithecal (d) Cell with chloroplasts (e) Ventral view with plate ornamentation (f) Dorsal (g) Hypothecal with single antapical plate (Thompson 1950, Figs. 73–79).

Figure 73. *Kansodinium ambiguum.* Overall view of cell (LM).

Figure 74. Ventral hypotheca and sulcus of empty cell (LM).

Figure 75. Ventral; note cingulum without displacement, sulcus not to antapex (SEM).

Figure 76. Epitheca, apical view; note apical pore and horseshoe-shaped ridge around it (Carty 2003, Fig. 7D).

Figure 77. Hypotheca, antapical view; note single antapical plate (SEM).

Plate formula: apical pore, pp, cp, 3′, 1a, 5″, 5‴, 1‴″; freshwater, photosynthetic.

One species reported from North America, *Kansodinium ambiguum*.

Kansodinium ambiguum

(Thompson) Carty and Cox 1986, 197.

BASIONYM: *Glenodinium ambiguum* Thompson 1950, 294, Figs. 73–79.

SYNONYM: *Diplopsalis ambiguum* (Thompson) Bourrelly 1970.

Round cell, brownish; eyespot present; one antapical plate. **Figs. 72–77**

EXTERNAL FEATURES: Plate formula: apical pore, pp, cp, 3′, 1a, 5″, 5‴, 1‴″; round cell, epitheca hemispherical; hypotheca flattened antapically; dorsoventral compression; large 3′ plate; small, four-sided 1a on left dorsal surface; single large antapical plate; cingulum median with slight displacement; sulcus not entering epitheca, sides narrow and parallel, not reaching antapex; apical pore surrounded by keyhole-shaped ridge; plates with light reticulate to striate ornamentation.

INTERNAL FEATURES: Golden to dark brown chloroplasts; distinct red eyespot.

SIZE: 31–42 µm long by 27–40 µm wide, 27–31 µm thick.

ECOLOGICAL NOTES: Found in plankton, can form bloom (Carty 1986).

LOCATION REFERENCE: KS (Thompson 1950), TX (Carty 1986).

Lophodinium

Lemmermann 1910, 629, 637, Figs. 18–33.

TYPE SPECIES: *Lophodinium polylophum* (Daday) Lemmermann 1910, 637.

Cell with theca divided by longitudinal ridges, apical slit.

Theca composed of numerous hexagonal plates (difficult to see) in longitudinal rows, their margins found in the valleys between ridges; ridges double. Freshwater, photosynthetic, with eyespot. One species known from North America, *Lophodinium polylophum*.

Lophodinium polylophum

(Daday) Lemmermann 1910, 637.

BASIONYM: *Glenodinium polylophum* Daday 1905, 23–25 Taf. 1, Figs. 18–22.

Longitudinal ridges from apex and antapex to cingulum. **Figs. 78–84; CP6(6)**

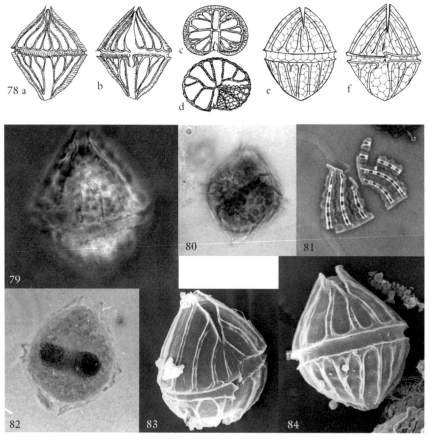

Figure 78. *Lophodinium polylophum.* (a) Dorsal view; note longitudinal ridges and apical slit (b) Ventral (c) Apical (Daday 1905, Figs. 18, 20, 21) (d) Epitheca with plates (Tafall 1942, Fig. 7) (e) Dorsal view showing relationship between ridges and hexagonal plates (f) Ventral view (orig.).

Figure 79. *Lophodinium polylophum.* Note opening at apex, vertical ridges (LM).

Figure 80. Cell dividing; note chloroplasts, ridges (LM) (Carty and Cox 1985).

Figure 81. Epitheca flattened to show ridges composed of hexagonal plates (Carty and Cox 1985).

Figure 82. Central, U-shaped nucleus (LM) (Carty and Cox 1985).

Figure 83. Ventral; note apical gap, ridges, and zigzag of plate margins between ridges (SEM) (Carty and Cox 1985).

Figure 84. Dorsal; note apical gap, ridges (SEM) (Carty 2003, Fig. 8F).

EXTERNAL FEATURES: Cell shaped like two cones, epitheca and hypotheca tapering from the median cingulum; apical ridge extends across top of cell from midventral to middorsal position, is not continuous with the sulcus and separates forming a slit; cingulum slightly displaced; sulcus extends to antapex with a list on the left side (Carty and Cox 1985).

INTERNAL FEATURES: Numerous golden chloroplasts; prominent eyespot in the sulcus; nucleus median.

SIZE: 42–44 µm long by 31–41 µm wide; dividing cells larger, 52–62 µm by 42–54 µm.

LOCATION REFERENCE: I have collected this species twice from the same location, a small pond in College Station, TX, in June, in large numbers (Carty and Cox 1985). The original report is from Paraguay (Daday 1905), one report from Mexico (Osorio-Tafall 1942), one from Bolivia (Iltis and Couté 1984). Recently collected from Florida (2007, C. Delwitch pers. comm.).

Naiadinium

Carty herein.

TYPE SPECIES: *Naiadinium polonicum* (Wołoszyńska) Carty herein.

Rhomboid-ovate cell with an apical pore, 4′, 1-2a, 7″, 5C, 5‴, 2⁗.

Freshwater, planktonic, large 1′ plate reaches apex, anterior intercalary plate(s) in middorsal position between the 3′ and 4″ plates, antapical plates unequal, with 1⁗ smaller than the 2⁗.

Key to species of *Naiadinium*

1a. One apical intercalary plate... *N. polonicum*
1b. Two apical intercalary plates .... *N. biscutelliforme*

Naiadinium biscutelliforme

(Thompson) Carty herein.

BASIONYM: *Peridinium polonicum* var. *biscutelliforme* Thompson 1950, 293–294, Figs. 35–40.

Deeply concave ventral face, single prominent antapical spine, 2a. **PL18(1); Figs. 85–87; CP7(1)**

EXTERNAL FEATURES: Plate formula: 4′, 2a, 7″, 5C, 5‴, 2⁗; epitheca with large 1′ plate reaching apex, two intercalary plates middorsal over the 4″ plate; ventral face deeply concave, dorsal side convex; cingulum displaced one cingulum width; sulcus does not enter epitheca, has parallel sides to the antapex; a single spine extends from the ventral junction of the two antapical plates; antapical plates are unequal, the 1⁗ smaller than the 2⁗.

INTERNAL FEATURES: Photosynthetic, with small disk-shaped, golden plastids, stigma in sulcus.

SIZE: 33–54 µm long by 30–51 µm wide.

ECOLOGICAL NOTES: Common in large bodies of water.

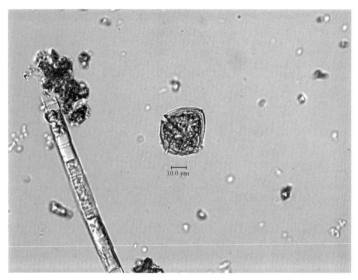

Figure 85. *Naiadinium biscutelliforme*. Ventral, single heavy posterior spine (LM).

Figure 86. Ventral; note large 1′ plate, single heavy posterior spine (*arrow*), ridges associated with apical pore.

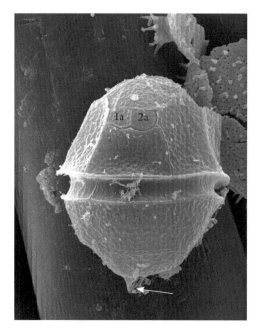

Figure 87. Dorsal; note two small intercalary plates, posterior spine (*arrow*) (SEM).

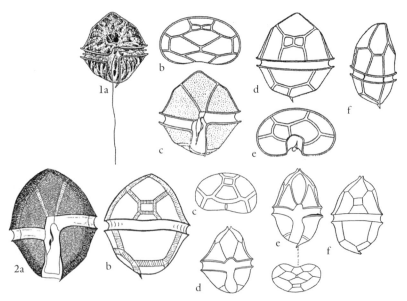

Plate 18. Species of *Naiadinium*. 1. *N. biscutelliforme* (a) Ventral, showing chloroplasts, eyespot, and flagella (b) Epitheca (c)Ventral with plates (d) Dorsal (e) Hypotheca (f) Side view (Thompson 1950, Figs. 35–40). 2. *N. polonicum*. (a) Ventral (b) Dorsal; note single apical intercalary plate and hypothecal spine (c) Epitheca (Wołoszyńska 1916, Pl. 12, Figs. 1, 2, 6) (d, e) Ventral (f) Dorsal (Tiffany and Britton 1952, Pl. 86, Figs. 1000–1002, with permission).

Naiadinium polonicum

(Wołoszyńska) Carty comb. nov.

BASIONYM: *Peridinium polonicum* Wołoszyńska 1916, 271, Taf. 12, Figs. 1–10.

SYNONYMS: *Glenodinium gymnodinium* Penard 1891, 54; *Peridiniopsis gymnodinium* (Penard) Bourrelly 1968a; *Peridiniopsis polonicum* (Wołoszyńska) Bourrelly 1968a.

Deeply concave, single prominent antapical spine, 1a. **PL18(2)**

EXTERNAL FEATURES: Cell oval; ventral face concave; single antapical spine on left side; cingulum median with slight displacement; sulcus not entering epitheca, widely spreading in hypotheca to antapex; irregular punctate ornamentation: dorsal plates distinctive: as illustrated, the 4′ is remote from the apex and the 1a is short and narrow and runs the length of the 4″ plate (Popovský 1970).

SIZE: 38–40 µm long by 35–36 µm wide, 15–16 µm deep.

ECOLOGICAL NOTES: Many reports due to EPA surveys 1977–1979 that differentiated the 1a and 2a forms, I have collected only *N. biscutelliforme*.

Palatinus

Craveiro, Calado, Daugbjerg, and Moestrup 2009, 1177.

TYPE SPECIES: *P. apiculatus* (Ehrenb.) Craveiro, Calado, Daugbjerg, and Moestrup 2009, 1178.

Sulcus strongly skewed, two apical intercalary plates, no apical pore.

Plate pattern: 4′, 2a, 7″, 6C, 5S, 5‴, 2⁗; no apical pore; plates smooth to granular without ridges; sulcus strongly skewed; chloroplast lobed; eyespot present; ecdysing through antapical-postcingular area (Craveiro et al. 2009). This genus was removed from *Peridinium,* group Palatinum, based on differences in plate pattern compared with *Peridinium cinctum* (the type species), and molecular and morphological differences (Craveiro et al. 2009).

Key to species of *Palatinus*

1a. Epithecal plate margins lack extensions, sutures between plates heavily striated ... .*P. pseudolaevis*
1b. Epithecal plate margins with extensions, sutures between plates lightly striated 2
2a. Apical plates strongly asymmetrical, hypothecal plates heavily spined............... ....*P. apiculatus*
2b. Apical plates somewhat asymmetrical, hypothecal plates lightly spined ....*P. apiculatus* var. *laevis*

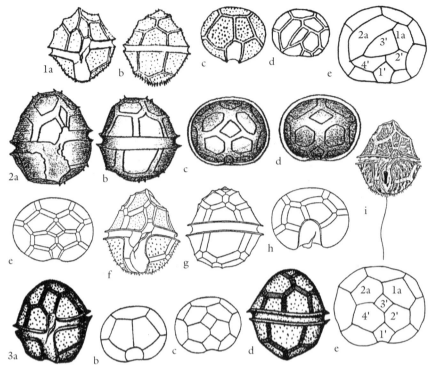

Plate 19. Species of *Palatinus*. 1. *P. apiculatus*. (a) Ventral; note ridges at plate boundaries (b) Dorsal; note heavy spines at antapex (c) Antapical (d) Apical (Lemmermann 1910, Figs. 5–8, 658 as *Peridinium marssonii*) (e) Epitheca with plate tabulation; note strong asymmetry of apical plates (Popovský and Pfiester 1990 as *Peridinium palatinum*). 2. *P. apiculatus* var. *laevis*; note plate margin extensions. (a) Ventral (b) Dorsal (c, d) Epitheca (Huitfeldt-Kaas 1900, Figs. 1–4 as *Peridinium laeve*) (e) Epitheca (f) Ventral view; note skewed sulcus, small spines on plates of hypotheca (g) Dorsal (h) Hypotheca (i) Ventral view; note eyespot in sulcus and chloroplast arrangement (Thompson 1950, Figs. 68–72 as *Peridinium palatinum*). 3. *P. pseudolaevis*. (a) Ventral (b) Antapical (c) Apical (d) Dorsal; note heavily striated intercalary bands (Lemmermann 1910, Figs. 5–8, 658 as *Peridinium marssonii*) (e) Plate pattern.

Palatinus apiculatus

(Ehrenberg) Craveiro, Calado, Daugbjerg, and Moestrup 2009, 1178.

BASIONYM: *Glenodinium apiculatum* Ehrenberg 1838, 258, Pl. 22, Fig. 24.

SYNONYMS: *Peridinium palatinum* Lauterborn 1896, 17–18, description only; *Peridinium marsonii* Lemmermann 1900b, 28, description only.

Hypothecal plates and plate margins with short thick spines; sulcus strongly skewed; two apical intercalary plates; no apical pore; 3′ and 2a elongate. **PL19(1)**

EXTERNAL FEATURES: Plate pattern: 4', 2a, 7″, 6C, 5S, 5‴, 2⁗; cell ovoid, epitheca conical, hypotheca rounded; plates with light granular ornamentation; apical plates strongly asymmetrical with small, narrow 1' plate not reaching the apex formed by plates 2', 3', and 4'; 2a elongate and touching 4″, 5″, and 6″; apical intercalary plates look asymmetrical in apical view, less so in dorsal view; 2a elongate; epithecal plate margins extended; cingulum slightly displaced, skewed sulcus barely enters epitheca, reaches antapex with parallel sides; hypothecal plate margins extended into spines, hypothecal plates with small spines; sutures between plates lightly striated.

INTERNAL FEATURES: Nucleus median; central pyrenoid with radiating chloroplast lobes, eyespot present in sulcus (Craveiro et al. 2009).

SIZE: Cell 32–48 μm long by 28–42 μm wide.

ECOLOGICAL NOTES: Collected by Prescott (1927) in Iowa as *Peridinium marsonii* in July from a quarry pond.

LOCATION REFERENCE: ON (Duthie and Socha 1976 as *Peridinium palatinum*), IA (Prescott 1927 as *Peridinium marsonii*), Michoacán (Tafall 1941).

Palatinus apiculatus var. *laevis*

(Huitfeldt-Kaas) Craveiro, Calado, Daugbjerg, and Moestrup 2009, 1178.

BASIONYM: *Peridinium laeve* Huitfeldt-Kaas 1900, 4, Figs. 1–5.

SYNONYM: *Peridinium palatinum* fa *laeve* (Huitfeldt-Kaas) Lefèvre 1932.

Hypothecal plates fringed with small spines, no apical pore, two intercalary plates. **PL19(2)**

EXTERNAL FEATURES: Plate pattern: 4', 2a, 7″, 6C, 5S, 5‴, 2⁗; species with elevated epithecal plate margins, spiny hypothecal plate margins; 3' plate not elongate but sides almost equal; apical intercalary plates more symmetric; spines of hypothecal plates fewer and smaller; sutures between plates lightly striated, epithecal plates smooth with pores. Recognized in illustrations by unstriated sutures, size and shape of 3' plate, and more symmetrical arrangement of apical plates. Cysts smooth. (Thompson 1950).

INTERNAL FEATURES: Narrow eyespot (Thompson 1950).

SIZE: 34–53 μm long by 28–46 μm wide (Thompson 1950).

ECOLOGICAL NOTES: Collected by Thompson (1950) during cool seasons only (November to April), sometimes as a bloom, absent in the summer, in Kansas.

LOCATION REFERENCE: ON (Duthie and Socha 1976 as *Peridinium palatinum* var. *laeve*), KS (Thompson 1950).

Palatinus pseudolaevis

(Lefèvre) Craveiro, Calado, Daugbjerg, and Moestrup 2009, 1178.

BASIONYM: *Peridinium pseudolaeve* Lefèvre 1925 341, Pl. 11, Figs. 6–9.

No apical pore, two apical intercalary plates, epithecal plates lack extensions. **PL19(3)**

EXTERNAL FEATURES: Plate pattern: 4', 2a, 7", 6C, 5S, 5''', 2''''; no apical pore, epithecal plate margins lack extensions and hypothecal plates lack heavy spines; sutures between plates striated; apical and apical intercalary plates symmetrical and similar in size; strongly skewed sulcus with slight penetration into epitheca, barely or not reaching antapex; slight dorsoventral compression.

SIZE: 28–37 μm long by 25–35 μm wide.

LOCATION REFERENCE: Labrador (Duthie et al. 1976), ON (Duthie and Socha 1976 as *Peridinium pseudolaeve*).

Parvodinium

Carty 2008.

TYPE SPECIES: *Parvodinium umbonatum* (Stein) Carty 2008, 106.

Small cells, theca obvious, apical pore, species may have spines.

Plate pattern: apical pore, canal plate, 4', 2a, 7", C6, 5''', 2''''; small, ovoid to pentagonal freshwater cell, most species are photosynthetic with yellow-gold plastids, most species with sulcus penetrating the epitheca and spreading to the antapex; 3' and 4" plates may be in *conjunctum, contactum,* or *remotum* positions. In some species the cingulum is submedian and the hypotheca is smaller than the epitheca; the cingulum is wide, contributing up to 18% of the total length. Care should be taken during collecting of samples; cells pass through wide-gauge nets, need 10 μm net or settle or centrifuge whole water.

Key to species of *Parvodinium*

1a. Cell nonphotosynthetic, single slender antapical spine .... *P. goslaviense*
1b. Cell yellow gold 2
2a. Posterior spines present 3
2b. Posterior spines lacking 6
3a. Two heavy posterior spines .... *P. deflandrei*
3b. Smaller spines, few to many 4
4a. Cell pentangular 5
4b. Cell with rounded epitheca .... *P. umbonatum* (in part)
5a. Thick short spines, 3' and 4" always contactum .... *P. africanum*
5b. Small cell, small spines, 3' and 4" various .... *P. inconspicuum* (in part)

Parvodinium africanum

(Lemmermann) Carty 2008, 106.

BASIONYM: *Peridinium africanum* Lemmermann ex West 1907, 188, Pl. 9, 1a–e.

SYNONYMS: *Peridinium intermedium* Playfair 1919; *Peridinium africanum* var. *intermedium* (Playfair) Lefèvre 1932.

Heavy posterior spines, apical intercalary plates diamond-shaped. **PL20(1); Fig. 88**

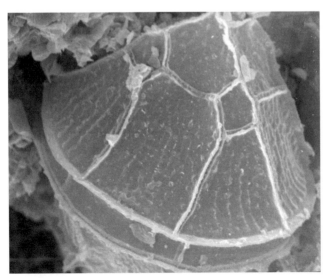

Figure 88. *Parvodinium africanum.* Note arrangement of dorsal epithecal plates, 3′ and 4″ *contactum*, apical intercalary plates small and similar.

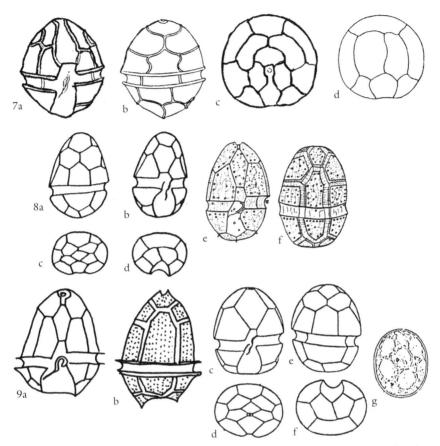

Plate 20. Species of *Parvodinium*. 1. *P. africanum*. (a) Dorsal; note configuration of 4″ plate and apical intercalaries (b) Ventral (c) Epithecal plate pattern with 3′ and 4″ contactum (Playfair 1919, text figs. 16a,b,c as *Peridinium intermedium*). 2. *P. belizensis*. (a) Ventral view; note wide cingulum (b) Dorsal with 3′ and 4″ remotum (c) Epitheca (d) Hypotheca with dissimilar antapical plates (redrawn from Carty and Wujek 2003, Fig. 14). 3. *P. centenniale*. (a) Dorsal (b) Ventral; note position of apical pore (c) Epitheca (d) Hypotheca (redrawn from Playfair 1919, text fig. 14). 4. *P. deflandrei*. (a) Dorsal; note two long posterior spines (b) Ventral; note extent of wide sulcus (c) Dorsal epitheca (d) Ventral with ornamentation (e) Hypotheca (f) Cyst (g) Epitheca (Lefèvre 1932, Figs. 543, 544, 546, 547) (h) Ventral (orig.). 5. *P. inconspicuum*. (a) Epitheca (b) Dorsal (c) Ventral (d) Hypotheca (Thompson 1947, Pl. Figs. 16–19) (e) Ventral; note wide cingulum, wide sulcus (orig.). 6. *P. lubieniense*. (a) Ventral (b) Dorsal (c) Hypotheca (d) Location of nucleus (Wołoszyńska 1916, Pl. 12, Figs. 21–24). 7. *P. morziense;* note wavy plate margins. (a) Ventral (b) Dorsal (c) Epitheca (d) Hypotheca (Lefèvre 1925 Figs. 1–4 as *Peridinium elegans*). 8. *P. pusillum*. (a) Dorsal (b) Ventral (c) Epitheca (d) Hypotheca (Forest 1954a Fig. 520) (e) Ventral view of young cell with ornamentation and sulcal plates (f) Dorsal of older cell with wide intercalary bands (orig.). 9. *P. umbonatum*. (a) Ventral (b) Dorsal; note shape of epitheca and proportion of epitheca to hypotheca (redrawn from Stein 1883 Figs. 1, 2) (c) Ventral (d) Epitheca (e) Dorsal (f) Hypotheca (g) Cyst (Thompson 1947, Plate 2, Figs. 20–24).

EXTERNAL FEATURES: Plate formula: apical pore, pp, cp, 4′, 2a, 7″, 5‴, 2⁗; horseshoe-shaped pore plate, cover plate, rectangular canal plate, 3′ and 4″ always conjunctum. This species, which was originally figured as 3′ and 4″ remotum, intercalary plates six sided, has been found only in the condition of 3′ and 4″ conjunctum with diamond-shaped intercalary plates in North America. The heavy posterior spines are distinctive. The cingulum is without displacement, sulcus penetrates the epitheca and expands widely in the hypotheca, reaching the antapex; hypothecal plates are flat to slightly concave, and the cell posterior is flat to concave; plates lightly reticulate; precingular plates may have vertical rows.

SIZE: 25–35 µm long by 23–27 µm wide.

Parvodinium belizensis

(Carty) Carty 2008, 106.

BASIONYM: *Peridinium belizensis* Carty in Carty and Wujek 2003, 137, Fig. 14.

Sulcus does not reach antapex, unequal antapical plates. **PL20(2); Figs. 89, 90**

EXTERNAL FEATURES: Plate formula: apical pore, pore plate, canal plate, 4′, 2a, 7″, S4, 5‴, 2⁗. Cingulum slightly displaced, sulcus penetrating epitheca, not reaching antapex, antapical plates of unequal size, the 1⁗ small and the 2⁗ large with small spines.

Figure 89. *Parvodinium belizensis.* Dorsal view, note 3′ and 4″ remotum, wide cingulum.

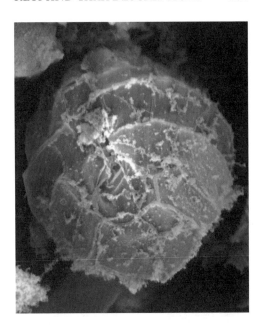

Figure 90. Note difference in sizes of antapical plates (SEM).

INTERNAL FEATURES: Plastids yellow.

SIZE: 12–16 μm long by 10–13 μm wide.

LOCATION REFERENCE: Belize (Carty and Wujek 2003).

Parvodinium centenniale

(Playfair) Carty 2008, 106.

BASIONYM: *Peridinium umbonatum* var. *centenniale* Playfair 1919, 806, Fig. 14.

Spherical shape, eccentric apical pore. **PL20(3); Figs. 91, 92**

EXTERNAL FEATURES: Plate formula: apical pore, pore plate, canal plate, 4', 2a, 7″, 5‴, 2⁗. Originally described from Centennial Park, Sydney. Popovský and Pfiester (1986, 1990), in consolidating the species, varieties, and forms, lumped most of the variation into *P. umbonatum*, but did retain five varieties, *centenniale* being one of them. The cingulum is submedian, dividing the cell into about two-thirds epitheca, one-third hypotheca, and has slight displacement. The sulcus penetrates the epitheca and reaches the antapex. The cells were illustrated as 3' and 4″ conjunctum, intercalary plates five sided, anatapical plates about equal in size.

SIZE: 29 μm long by 26 μm wide (Wailes 1934).

LOCATION REFERENCE: BC (Wailes 1934), Belize (Carty and Wujek 2003).

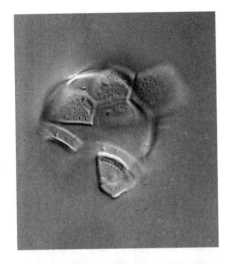

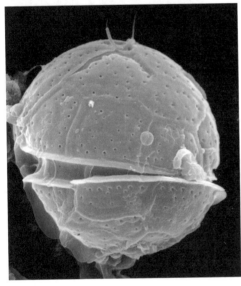

Figure 91. *Parvodinium centenniale*. Note hemispherical epitheca, epitheca losing plates (LM).

Figure 92. Dorsal (SEM).

Parvodinium deflandrei

(Lefèvre) Carty 2008, 106.

BASIONYM: *Peridinium deflandrei* Lefèvre 1927, 121.

Conical epitheca, two heavy posterior spines. **PL20(4)**

EXTERNAL FEATURES: Plate formula: apical pore, 4′, 2a, 7″, 5‴, 2⁗; conical epitheca, plates flat to slightly concave, linear punctate ornamentation, median cingulum without displacement. Sulcus enters

epitheca and is very broad in the hypotheca (only plates 1‴ and 5‴ are visible), 3′ and 4″ illustrated as conjunctum. Heavy spine at antapex and at base of 5‴ plate.

ECOLOGICAL NOTES: Capable of forming summer bloom (Canion and Ochs 2005).

Parvodinium goslaviense

(Wołoszyńska) Carty 2008, 106.

BASIONYM: *Peridinium goslaviense* Wołoszyńska 1916, 267, Taf. 10, Figs. 18–24.

Nonphotosynthetic, single spine at antapex. **PL15(3); Figs. 93, 94**

EXTERNAL FEATURES: Plate formula: apical pore, 4′, 2a, 7″, 5‴, 2⁗; ovate cell, slightly apiculate or with apical notch; 3′ and 4″ illustrated as contactum; shallow cingulum, median, with slight displacement; sulcus entering epitheca and descending with parallel sides to antapex; antapical plates about equal (2⁗ > 1⁗), and small compared to large 2‴, 3‴, and 4‴; punctate ornamentation; long, single spines at antapex. Although not all cingular plates are illustrated by Wołoszyńska (1916), there are four by the middorsal alignment, suggesting six cingular plates.

INTERNAL FEATURES: Nucleus is in the hypotheca, no plastids; may have colored bodies (red gold), especially in epitheca.

SIZE: 20–25 µm long by 16–18 µm wide.

ECOLOGICAL NOTES: Swims vigorously (Wołoszyńska 1916).

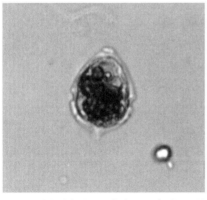
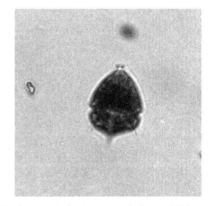

Figures 93, 94. *Parvodinium goslaviense.* Note single antapical spine, extended apex (LM).

Parvodinium inconspicuum

(Lemmermann) Carty 2008, 106.

BASIONYM: *Peridinium inconspicuum* Lemmermann 1899, 350.

Small, pentangular cell; common. **PL20(5); Figs. 95–96; CP7(2,3)**

EXTERNAL FEATURES: Plate formula: apical pore, pore plate, cover plate, canal plate, 4′, 2a, 7″, C6, 5‴,

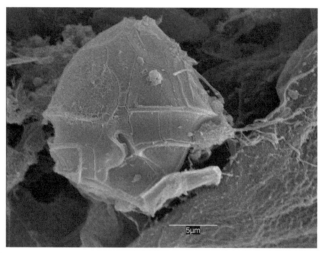

Figure 95. *Parvodinium inconspicuum.* Ventral view; note wide sulcus, small hypotheca (SEM).

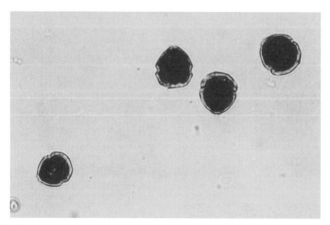

Figure 96. Cells stained with iodine; note pentagonal shape and flattened posterior (LM).

2''''; epitheca angular (plates flat or slightly convex), antapex flat, or concave. Cingulum wide, without displacement, median in position, though epitheca may be larger than hypotheca; sulcus penetrating the epitheca and then expands widely in the hypotheca to the antapex. The 3' and 4'' plates may be in conjunctum, contactum, or remotum positions; plates are lightly reticulate. Antapical plates cannot be seen in ventral view; they are equal in size and symmetrically placed on the hypotheca; they may have small spines.

INTERNAL FEATURES: Plastids are yellow gold, peripheral in location; there is a large nucleus in the hypotheca; eyespots seen in some specimens, lacking in others.

SIZE: 13–26 μm long by 11–23 μm wide.

ECOLOGICAL NOTES: Species was reported dominant (99% total biovolume) in small acidic (pH 4.6) lakes (Perez et al. 1994).

Parvodinium lubieniense

(Wołoszyńska) Carty 2008, 106.

BASIONYM: *Peridinium lubieniense* Wołoszyńska 1916, 272, Taf. 12, Figs. 21–24.

Ovoid cell, punctate ornamentation on antapical plates. **PL20(6)**

EXTERNAL FEATURES: Cell ovoid, dorsoventrally flattened; theca thin, making plates and ornamentation difficult to see (Lefèvre 1932); antapical plates with papillae (Wołoszyńska 1916).

INTERNAL FEATURES: Disk-shaped chloroplasts, central nucleus.

SIZE: 35–45 μm long by 30–32 μm wide.

ECOLOGICAL NOTES: Reported as rare from bogs in Labrador (Duthie et al. 1976).

LOCATION REFERENCE: Labrador (Duthie et al. 1976).

Parvodinium morzinense

(Lefèvre) Carty 2008, 106.

BASIONYM: *Peridinium morzinense* Lefèvre 1928, 137.

SYNONYM: *Peridinium elegans* Lefèvre 1925, 332, Figs. 1-6.

Plate margins curved. **PL20(7)**

EXTERNAL FEATURES: Plate margins curved, without ornamentation; oval cell lacking dorsoventral compression; sulcus barely enters epitheca and widens slightly in hypotheca, reaching antapex.

SIZE: 30–41 μm long by 26–35 μm wide.

LOCATION REFERENCE: Not reported from North America.

Parvodinium pusillum

(Penard) Carty 2008, 106.

BASIONYM: *Glenodinium pusillum* Penard 1891, 52–53, Pl. 4, Figs. 1–4.

Ovate, sulcus not to antapex. **PL20(8); Fig. 97**

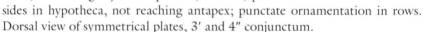

EXTERNAL FEATURES: Plate formula: apical pore, pore plate, canal plate, 4′, 2a, 7″, 5‴, 2″″; ovate to rounded cell with slight dorsoventral compression; cingulum submedian dividing the cell into a larger epitheca and smaller hypotheca, without displacement; sulcus slightly into epitheca, narrow, parallel sides in hypotheca, not reaching antapex; punctate ornamentation in rows. Dorsal view of symmetrical plates, 3′ and 4″ conjunctum.

INTERNAL FEATURES: Golden-brown plastids (Prescott 1951b).

SIZE: 18–24 μm long by 13–20 μm wide.

Figure 97. *Parvodinium pusillum.* Note rounded posterior, sulcus not to antapex (SEM).

Parvodinium umbonatum

(Stein) Carty 2008, 106.

BASIONYM: *Peridinium umbonatum* Stein 1883, Pl. 12, Figs. 1–8.

Oval cell, wide cingulum submedian, dividing cell into two-thirds epitheca, one-third hypotheca, wide sulcus to antapex. **PL20(9); Figs. 98, 99; CP7(4,5)**

EXTERNAL FEATURES: Plate pattern: apical pore, canal plate, 4′, 2a, 7″, C6, 5‴, 2″″, epitheca broadly rounded, hypotheca flat, or concave. The presence of an apical pore may be detected by a slight flattening at the apex of the cell. Sulcus penetrates the epitheca with a large Sa plate, there is an Sd flap over the sulcal pore, an Ss, Sm, and large, wedge-shaped Sp plate extending to the antapex. The 3′ and 4″ plates can be in conjunctum, contactum, or remotum positions; plates are lightly reticulate. Antapical plates cannot be seen in ventral view, they are equal in size and symmetrically placed on the hypotheca.

Figures 98, 99. *Parvodinium umbonatum.* Rounded epitheca, wide cingulum (LM, SEM).

INTERNAL FEATURES: Plastids are yellow gold, there is a large nucleus in the hypotheca, no eyespots seen.

SIZE: 16–28 µm long by 12–21 µm wide.

Peridiniopsis

Lemmermann 1904, 134.

TYPE SPECIES: *Peridiniopsis borgei* Lemmermann 1904, 134, Taf. 1, Figs. 1–5.

For a discussion of *Peridiniopsis,* see "Taxonomy" in the introduction.

Bourrelly (1968a,b), in reviewing the problem of the genus *Glenodinium,* chose *Peridiniopsis* as a more defined substitute. He cited Schiller's description of 3-5′, 0-1a, 6-7″, 5‴, 2″″ for the genus. Species were arranged in categories based on epithecal arrangement, for example, 3′-1a-6″, 4′-0a-6″, 5′-0a-6″, 3′-1a-7″, 4′-0a-7″, 4′-1a-7″, 5′-0a-7″, and 5′-1a-7″. All species but *Peridiniopsis edax* have an apical pore and are photosynthetic. In this key, since plates are difficult to see, other, more obvious morphological characters will be used first.

Key to species of *Peridiniopsis*

1a. Cell not photosynthetic, no apical pore ....*P. edax*
1b. Cell with yellow-gold plastids, with apical pore...2
2a. Cell with hypothecal spines 3
2b. Cell lacks hypothecal spines..7
3a. Slender spines only on antapical plates.......................................*P. lindemanii*
3b. Spines otherwise 4
4a. Numerous spines associated with sutures..................................... *P. elpatiewskyi*
4b. Single spines associated with hypothecal plates..5
5a. 4″ plate four sided...*P. quadridens* (a)
5b. 4″ plate five sided 6

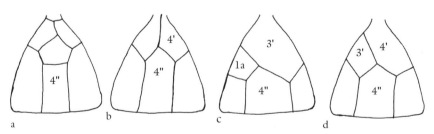

Figure 100. Dorsal plate patterns. (a) *Peridiniopsis quadridens,* 5′, 1a, 7″, 1a plate centered over four-sided 4″ (b) *Peridiniopsis thompsonii,* 5′, 7″; 3′ and 4′ symmetrical over the 4″ (c) *Peridiniopsis cunningtonii,* 4′, 1a, 6″; 1a and 3′ suture skewed (d) *Peridiniopsis cunningtonii,* 5′, 0a, 6″; 3′ and 4′ suture skewed.

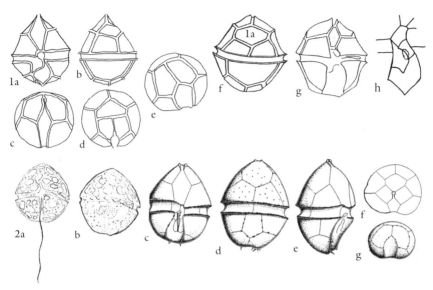

Plate 21. Ovoid-shaped species of *Peridiniopsis*. 1. *P. borgei* (3′ 1a 6″). (a) Ventral (b) Dorsal (c) Ventral epitheca (d) Apical (e) Antapical (Lemmermann 1904, Taf 1, Figs. 1–5) (f) Dorsal; note large 1a plate in central position (g) Ventral (Prescott 1951b, Pl. 90, Figs. 8–9 as *Glenodinium borgei*) (h) Sulcal plates (based on Calado and Moestrup 2002, Figs. 4, 5). 2. *P. penardii* (4′ 0a 6″). (a) Ventral (b) Dorsal (Penard 1891, Taf 3, Figs. 14, 15) (c) Ventral (d) Dorsal (e) Side (f) Epitheca (g) Hypotheca (Javornický 1971, Figs. 3–6, 8, 9) (c–g reprinted with permission from the Phycological Society of America).

6a. 4″ middorsal, 7 precingular plates..*P. thompsonii* (b)
6b. 4″ skewed, 6 precingular plates..*P. cunningtonii* (c, d)
7a. Cell round, nucleus median or in epitheca, eyespot present............................8
7b. Cell ovoid, nucleus in hypotheca 9
8a. 1a pentagonal, sulcus narrow.. *P. oculatum*
8b. 1a four sided, sulcus spreading .... *P. kulczynskii*
9a. Large 3′ plate fills dorsal epitheca, no "a" plate..............................*P. penardii*
9b. 1a plate fills center of dorsal epitheca..*P. borgei*

Peridiniopsis borgei

Lemmermann 1904, 134, Taf. I, Fig 1–5.

SYNONYM: *Glenodinium borgei* (Lemm) Schiller 1937.

Large 1a six sided across dorsal epitheca. **PL21(1)**

EXTERNAL FEATURES: Plate pattern: apical pore, 3′, 1a, 6″, C6, S5, 5‴, 2″″, with broad, six-sided 1a plate across dorsal epitheca. Cell ovoid, round in cross section; cingulum not displaced; sulcus not entering epitheca and extending to antapex, the Sd plate flaplike and the Sp concave (Calado and Moestrup 2002).

INTERNAL FEATURES: Nucleus in hypotheca, chloroplast lobes peripheral; may contain accumulation bodies in lower portion of epitheca, oil droplets in the epitheca, and starch grains in the hypotheca; eyespot with chloroplast may not be obvious; presence of a peduncle suggests the possibility of heterotrophic feeding (Calado and Moestrup 2002).

SIZE: 35 μm diameter, 40 μm long (Prescott 1951b).

ECOLOGICAL NOTES: Found in the tycoplankton (Prescott 1951b), and a reddish bloom on a lake (NC, August, Whitford and Schumacher 1984).

Peridiniopsis cunningtonii

Lemmermann in West 1907, 189, Pl. 9, Fig. 2.

3′/1a plate skewed to the left of the center dorsal position. **PL22(1); Fig. 101**

EXTERNAL FEATURES: Ovate cell with plate patterns 5′, 0a, 6″ (or 4′, 1a, 6″), C6, 5‴, 2⁗, spines on plates 1‴, 2‴, 4‴, 5‴, 1⁗, 2⁗. Though this taxon has two different plate patterns, the latter considered a

Figure 101. *Peridiniopsis cunningtonii* (epitheca only). Note skewed arrangement of dorsal plates, extended apex, and spine.

remotum event, it remains a pattern not found elsewhere. Viewing the cell from above, it can be imagined the 3′ plate being slightly squeezed from the apex, making it technically a 1a plate. The cingulum is median, without displacement; sulcus penetrates the epitheca at least one cingulum width and spreads in the hypotheca without reaching the antapex; antapical plates equal in size.

Original illustrations show both an apical 5′-0a view and a dorsal 4′-1a (recognized by the skewed position of the 1a plate) view. Prescott (1951b) illustrated *Peridiniopsis cunningtonii* with the skewed back, size 24–30 μm long by 20–25 μm wide, but called it *Glenodinium quadridens*.

INTERNAL FEATURES: Golden chloroplasts.

SIZE: 31–38 μm long by 27–31 μm wide (West 1907), cells found in OH were 27–28 μm long.

ECOLOGICAL NOTES: Tycoplankter (Prescott 1951b).

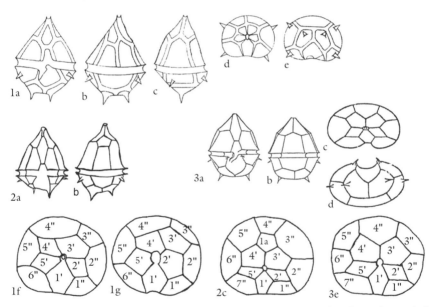

Plate 22. Ovate species of *Peridiniopsis* with single spines on postcingular and antapical plates. 1. *P. cunningtonii* (5′ 0a 6″, or 4′ 1a 6″). (a) Ventral (b, c) Dorsal showing 4′ 1a 6″ (d) Epitheca showing 5′ 0a 6″ (e) Hypotheca (Lemmermann in West 1907, Pl. 9, Figs. 2a–e) (f) 5′ 0a 6″ (g) 4′ 1a 6″. 2. *P. quadridens* (5′ 1a 7″). (a) Ventral (b) Dorsal (Stein 1883, Taf 11, Figs. 4, 6) (c) 5′ 1a 7″. 3. *P. thompsonii* (5′ 0a 7″). (a) Dorsal (b) Ventral (c) Epitheca (d) Hypotheca (Thompson 1947, Pl. 2, Figs. 12–15, as *Glenodinium quadridens*) (e) 5′ 0a 7″.

Peridiniopsis edax

(Schilling) Bourrellly 1968a, 9.

BASIONYM: *Glenodinium edax* Schilling 1891c, 206, Pl. 10, Figs. 23–24.

Nonphotosynthetic, no apical pore, sulcal list on right postcingular plate. **PL15(4)**

EXTERNAL FEATURES: Plate pattern: no apical pore, 3′, 1a, 6″, C6, 5‴, 2″″; roundish cells; cingulum median without displacement; sulcus not penetrating epitheca, with parallel sides, tapering to antapex; epithecal plates symmetrical, the second and third apical plates about equal in size, the 1a diamond-shaped in middorsal position; sulcal wing on right postcingular plate (5‴); no hypothecal spines.

INTERNAL FEATURES: Nonphotosynthetic, no eyespot, large nucleus in hypotheca, cytosol granular with refractive bodies.

SIZE: 15–27 μm long by 12–24 μm wide (Thompson 1950).

ECOLOGICAL NOTES: Many reports due to EPA surveys 1977–1979.

COMMENTS: Calado (2011) synonymized *Tyrannodinium berolinense* with *Peridiniopsis edax* on the basis of similar shapes, and on both being heterotrophic and lacking an eyespot. The work of Thompson (1950) showed different plate patterns, different apical symmetry, and only *T. berolinense* with an apical pore and antapical spines.

Peridiniopsis elpatiewskyi

(Ostenfeld) Bourrelly 1968a, 9 (also see B. Meyer and Elbrächter 1996).

BASIONYM: *Peridinium umbonatum* var. *elpatiewskyi* Ostenfeld 1907, 391, Pl. 9, Figs. 9–12.

SYNONYMS: *Peridinium elpatiewskyi* Lemmermann 1910, Figs. 20–24; *Peridinium marchicum* Lemmermann 1910, Figs. 16–19; *Peridinium marchicum* Lemm. var. *simplex* Wołoszyńska 1916.

Hypothecal plate margins extended into spines. **PL23(1); Figs. 102–104**

EXTERNAL FEATURES: Plate pattern: pp, cp, 4′, 0a, 7″, 6C, 4S, 5‴, 2″″; the 4″ four sided with the third apical above it; cell pentagonal; heavy with hypothecal spines; cingulum without displacement; sulcus entering epitheca and widely expanding in hypotheca to the antapex; reticulate or linear ornamentation; heavy spines on margins of antapical plates; base of cell flat

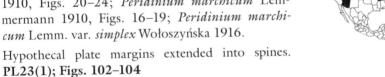

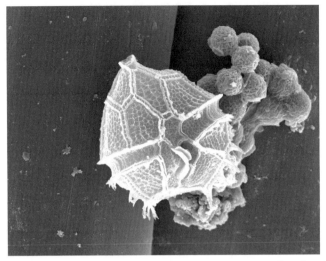

Figure 102. *Peridiniopsis elpatiewskyi.* Ventral view; note spines extending from hypothecal plate margins.

Figure 103. Side view; note spines and flattened posterior.

to concave. Thompson (1947) illustrated cells with spines extending from postcingular plates 1‴, 2‴, 4‴, and 5‴, but I have collected the species without such spines.

INTERNAL FEATURES: Chloroplasts yellow–golden brown; red eyespot left of sulcal region.

SIZE: 27–35 μm long by 30–37 μm wide.

Figure 104. Epitheca showing 4' 0a configuration.

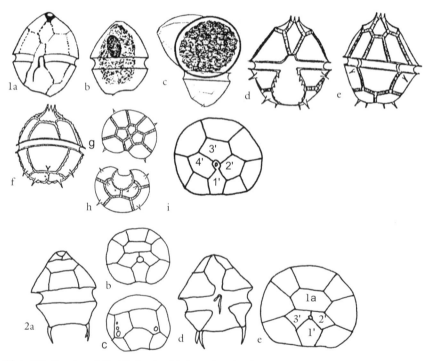

Plate 23. Species of *Peridiniopsis* with hypothecal spines. 1. *P. elpatiewskyi* (4' 0a 7″). (a) Ventral (b) Dorsal (c) Cell ecdysing (Ostenfeld 1907, Taf 9, Figs. 9, 11, 12) (d) Ventral (e) Dorsal (f) Dorsal hypotheca (g) Apical (h) Antapical (Thompson 1947, Pl. 2, Figs. 7–11) (i) Epithecal plate pattern (after Popovský and Pfiester 1990, Fig. 15). 2. *P. lindemanii* (3' 1a 7″). (a) Dorsal (b) Epitheca (c) Hypotheca (d) Ventral (Lefèvre 1932, Figs. 752–755) (e) Epithecal plate pattern (after Popovský and Pfiester 1990, Fig. 14).

ECOLOGICAL NOTES: Fair number seen in a reservoir (June, OH), also seen in large lake (August, NY), and reservoir (September–October, MD, Thompson 1947)

Peridiniopsis kulczyński

(Wołoszyńska) Bourrelly 1968a, 8.

BASIONYM: *Peridinium kulczyński* Wołoszyńska 1916, 272–273, Taf. 12, Figs. 25–31.

SYNONYM: *Glenodinium kulczyński* (Wołoszyńska) Schiller 1937.

Sulcus spreading to antapex; small 1a plate four-sided, diamond-shaped, central dorsal position. **PL24(1)**

EXTERNAL FEATURES: Plate pattern: apical pore, 3′, 1a, 6″, C6, 5‴, 2⁗, cell epitheca angularly round, apical plates 2′ and 3′ large, symmetrical. 1a diamond-shaped, in central dorsal position; hypotheca with flattened bottom, median cingulum without

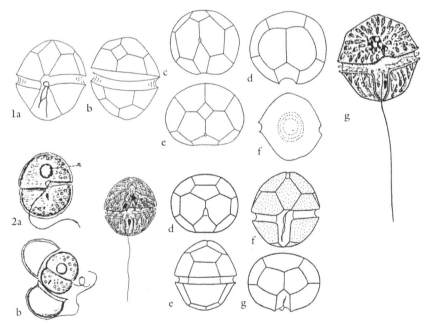

Plate 24. Round species of *Peridiniopsis* with eyespot, no spines. 1. *P. kulczynskii* (3′ 1a 6″). (a) Ventral with spreading sulcus (b) Dorsal (c) Epitheca (d) Hypotheca (e) Epitheca (f) Location of central nucleus (Wołoszyńska 1916, Taf 12, Figs. 25–31) (g) Ventral with radial chloroplasts and eyespot; note parallel sides to sulcus (Thompson 1950, Fig. 25). 2. *P. oculatum* (3′ 1a 7″). (a) Ventral (b) Cell ecdysing (Stein 1883, Taf 3, Figs. 5–6) (c) Ventral (d) Epitheca (e) Dorsal (f) Ventral (g) Hypotheca (Thompson 1950, Figs. 41–45).

displacement, sulcus not penetrating epitheca, with parallel sides in hypotheca, tapering to antapex; thin plates, antapical plates equal in size.

INTERNAL FEATURES: Plastids golden, parietal, appearing to radiate from center; small eyespot in sulcus, nucleus median (Thompson 1950).

SIZE: 28–48 µm long by 23–36 µm wide.

Peridiniopsis lindemanii

(Lefèvre) Bourrelly 1968a, 9.

BASIONYM: *Peridinium lindemanii* Lefèvre 1927, 121.

Small pentagonal cell with slender antapical spines. **PL23(2)**

EXTERNAL FEATURES: Small pentagonal cell; plate pattern: apical pore, 3′, 1a, 7″, 5‴, 2⁗, 1a plate six sided; slight dorsoventral compression; wide cingulum divides a larger epitheca from a smaller hypotheca; sulcus slightly penetrates epitheca and widens greatly in hypotheca, reaching antapex; plates without ornamentation except for a variable number of slender spines on antapical plates (Lefèvre 1932).

SIZE: 10–12 µm long by 9–11 µm wide.

LOCATION REFERENCE: Not reported from North America.

Peridiniopsis oculatum

(Stein) Bourrelly 1968a, 9.

BASIONYM: *Glenodinium oculatum* Stein 1883, Taf. 3, Figs. 5–7.

1a plate pentagonal; sulcus narrow with parallel sides. **PL24(2)**

EXTERNAL FEATURES: Plate pattern: apical pore, 3′, 1a, 7″, 5‴, 2⁗; roundish cell; slight dorsoventral compression; apex rounded; median cingulum with slight displacement; sulcus not penetrating epitheca, with straight sides in hypotheca, tapering to antapex; antapical plates equal in size; theca delicate, difficult to see until cell ecdyses, smooth to slightly punctate (Thompson 1950).

INTERNAL FEATURES: Rod-shaped eyespot in sulcus, golden parietal chloroplasts, nucleus in epitheca.

SIZE: 23–36 µm long by 21–36 µm wide (Thompson 1950).

ECOLOGICAL NOTES: Many reports due to EPA surveys 1977–1979.

Peridiniopsis penardii

(Lemmermann) Bourrelly 1968a, 9.

BASIONYM: *Glenodinium penardii* Lemmermann 1900a, 117, named for the cell called *Glenodinium cinctum* by Penard 1891, 52, Taf. 3, Figs. 14–21.

Large 3′ plate fills dorsal epitheca. **PL21(2)**

EXTERNAL FEATURES: Plate pattern: distinct apical pore, 4′, 0a, 6″, 5‴, 2⁗; cell round, sometimes slightly lobed in the posterior; theca thin; cingulum median without displacement; sulcus not into epitheca, with slightly spreading sides to antapex; hypotheca dorsoventrally compressed.

INTERNAL FEATURES: Numerous red-brown to yellow-brown chloroplasts; nucleus in hypotheca, may have red bodies on ventral face (Javornický 1971).

SIZE: 30 µm long by 28 µm wide (Penard 1891).

ECOLOGICAL NOTES: Found in bloom along inshore sites along lake (April, CA, Horne et al. 1972).

Peridiniopsis quadridens

(Stein) Bourrelly 1968a, 9.

BASIONYM: *Peridinium quadridens* Stein 1883, Taf. 11, Figs. 3–6.

SYNONYM: *Glenodinium quadridens* (Stein) Schiller 1937.

Pentagonal 1a in middorsal position, four spines on hypotheca, cell apex extended into a chimney. **PL22(2); Figs. 105–107; CP7(6)**

EXTERNAL FEATURES: Plate pattern: apical pore, 5′, 1a, 7″, C6, S5, 5‴, 2⁗; cell with conical, apiculate epitheca; rounded hypotheca; slightly dorsoventrally compressed; cingulum slightly displaced, dividing cells into a larger epitheca and smaller hypotheca; sulcus penetrating epitheca more than one cingulum width, expanding in the hypotheca to the antapex; large rectangular Sa plate, Sd flap over pore, Ss and Sm covered by Sd, large Sp, Sa, Sp, and Sd with reticulate ornamentation; dorsal epitheca with pentagonal 1a plate above four-sided 4″; hypotheca with spines on the second and fourth postcingulars (sometimes on the first and fifth also) and both antapicals; light reticulation on plates.

INTERNAL FEATURES: Eyespot (Prescott 1927), golden-brown chloroplasts.

SIZE: 30–40 µm long by 20–33 µm wide.

ECOLOGICAL NOTES: Graham et al. (2004) reported it as dominating the spring and fall plankton in Crystal Bog, WI. Many reports due to EPA surveys 1977–1979.

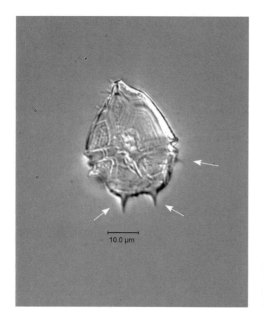

Figure 105. *Peridiniopsis quadridens.* Ventral view with antapical and postcingular spines (*arrows*) (courtesy of John D. Hall).

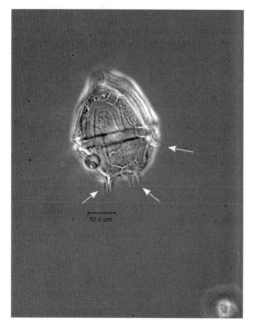

Figure 106. Dorsal view with four-sided 4″ plate (LM) (courtesy of John D. Hall).

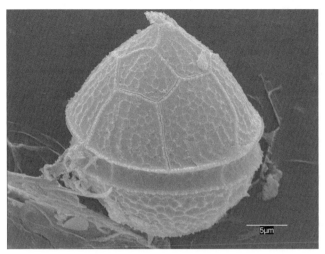

Figure 107. Dorsal epitheca with four-sided 4″ plate and central 1a plate (SEM).

COMMENTS: Three taxa sometimes mistaken for each other, *Peridiniopsis quadridens, Peridiniopsis cunningtonii,* and *Peridiniopsis thompsonii,* all have a rounded posterior with single spines emerging from the 2‴, 4‴, 1⁗ and 2⁗ plates **(Fig. 100, PL22).** They have different plate patterns: 5′, 1a, 7″ for *P. quadridens;* 5′, 0a, 6″ or 4′, 1a, 6″ for *P. cunningtonii;* and 5′, 0a, 7″ for *P. thompsonii,* but the differences may not be readily apparent. Epithecal dorsal views identify the taxon; *P. quadridens* has a large pentagonal 1a plate sitting on top of the square 4″ plate, *P. cunningtonii* has the 3′/1a plate skewing to the left from the center dorsal position, and *P. thompsonii* has a five-sided 4″ plate. A review of the literature in which authors have drawn what they see reveals errors in identification. For *P. quadridens,* Tiffany and Britton (1952) and Forest (1954a) identified *"Glenodinium quadridens"* but illustrated *P. thompsonii.* Popovský (1970) identified *Peridinium cunningtonii* but illustrated *Peridiniopsis quadridens.* Identifications without original drawings are suspect, but *P. quadridens* is the most common of the three species.

Peridiniopsis thompsonii

Bourrelly 1968a, 9.

Based on *Glenodinium quadridens* in Thompson 1947, 14, Pl. 2, Figs. 12–15.

Five-sided 4″ plate, strong dorsoventral compression. **PL22(3); Fig. 108**

EXTERNAL FEATURES: Plate pattern: apical pore, 5′, 0a, 7″, 5‴, 2⁗; ovoid cell; dorsoventrally compressed; single spines on hypothecal plates 2‴, 4‴, 1⁗, 2⁗; cingulum without displacement; sulcus

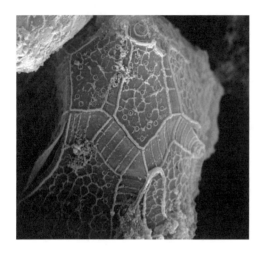

Figure 108. *Peridiniopsis thompsonii* with five apical plates surrounding apical pore.

entering epitheca about two cingulum widths, wide in upper hypotheca, narrowing to antapex; antapical plates about equal in size (Thompson 1947).

INTERNAL FEATURES: Tiffany and Britton (1952) report an eyespot; brownish golden chloroplasts (Thompson 1947).

SIZE: 31–38 μm long by 27–31 μm wide, 18.5 μm thick (Thompson 1947).

ECOLOGICAL NOTES: Found in a reservoir (August, Thompson 1947), and a large lake (August, OH), planktonic.

COMMENTS: Thompson (1947) identified cells as *Glenodinium quadridens,* as did Tiffany and Britton (1952), though both papers illustrated 5'-0a-7"; and Forest (1954a), though referring to Thompson 1947 and illustrating 5'-0a-7", called it *Glenodinium quadridens.* Eddy (1930) called it *Peridinium cunningtonii* but illustrated 5'-0a-7". Bourrelly (1968a) recognized that the taxon illustrated by Thompson was different and named the species for Thompson.

Peridinium

Ehrenberg 1830, 38.

TYPE SPECIES: *Peridinium cinctum* (O.F.M.) Ehrenberg 1830, 38.

Cell with thick thecal plates; pattern 4', 3a, 7", 5C, 5''', 2''''.

Peridinium is a genus of freshwater, predominantly photosynthetic species with thick thecal plates. Species are separated by the presence/absence of an apical pore, and the arrangement of apical and apical intercalary plates. The cingulum divides the cell into approximately equal halves; the epitheca may be slightly larger than the hypotheca. Antapical plates are usually about equal in size. All plates usually have ornamentation. Swimming pattern erratic, bumper car–like; check border outside coverslip for cells.

Lefèvre's 1932 monograph on the genus *Peridinium* divided the genus into two subgenera, Cleistoperidinium for species lacking an apical pore, and Poroperidinium for species with a pore. The subgenera were further divided into groups based on the number of apical intercalary plates and symmetry of the apical plates. While the pattern of the epitheca is definitive for most species, the epitheca can be clearly and completely seen only when the cell is empty and the epitheca dissociated from the hypotheca. Cells will usually come to rest on their ventral sides, which are more likely to be flat or concave than their dorsal sides. I have constructed the following key based on the epithecal dorsal view of species of *Peridinium* (they all look the same, platewise, in ventral view). Additional species of *Peridinium* can be found in Lefèvre (1932) and Starmach (1974).

Key to species of *Peridinium*

1a. No apical pore present 2
1b. Apical pore present..6

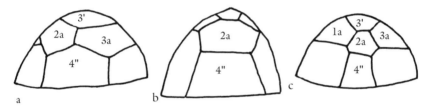

Figure 109. *Peridinium* dorsal epitheca plate patterns. (a) Cinctum pattern, 4″ five-sided (b) Willei pattern (also Bipes), 4″ four-sided (c) *Peridinium striolatum* with a small 4″ plate.

2a. 4″ plate 5 sided ....3 (Cinctum group) (a)
2b. 4″ plate 4 sided .... 4 (b)
3a. Cell strongly dorsoventrally compressed .... *P. raciborskii*
3b. Cell moderately compressed .... *P. cinctum*
3c. Cell round in cross section .... *P. gatunense*
4a. Small 2a plate atop small 4″ plate ....*P. striolatum* (c)
4b. Large 2a plate across large 4″ plate .... 5 (Willei group)
5a. Large 1′ to apex, strong compression .... *P. willei*
5b. Medium 1a, not to apex.. *P. volzii*
6a. 4″ plate 4 sided, posterior with 2 lobes.............................. 7 (Bipes group) (b)
6b. 4″ plate 6 sided 9

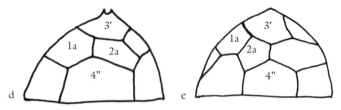

Figure 110. Five-sided 4″ patterns. (d) Gutwinskii pattern with 1a and 2a in contact (e) Lomnickii and Allorgei pattern with 2a and 3a in contact.

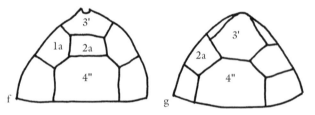

Figure 111. Six-sided 4″ patterns. (f) Gutwinskii pattern with 1a, 2a, and 3a in contact (g) Subsalsum pattern with 2a, 3′, and 3a in contact.

Peridinium achromaticum

Levander 1902, 49, Figs. 1–2.

Nonphotosynthetic, diamond shaped cell. **PL15(5)**

EXTERNAL FEATURES: Plate pattern: apical pore, 4′, 3a, 7″, 5‴, 2⁗, without chloroplasts, diamond-shaped cell, median cingulum without displacement, sulcus barely enters epitheca and extends to antapex. Posterior bilobed.

SIZE: 28–48 µm long by 24–40 µm wide (Popovský and Pfiester 1990).

ECOLOGICAL NOTES: Excluded by Lefèvre (1932), who considered it a saltwater species.

LOCATION REFERENCE: Not reported from North America.

Peridinium aciculiferum

Lemmermann 1900b, 28–29 (no illustration)

Ovoid cell, slim spines at three posterior sutures. **PL25(1)**

EXTERNAL FEATURES: Plate pattern: apical pore, 4′, 3a, 7″, C6, 5‴, 2⁗; cell ovate, epitheca tapered, hypotheca rounded, some dorsoventral compression; cingulum median without displacement, with six plates; sulcus with slight penetration into epitheca, reaches antapex; three long slim spines at antapex. Group Lomnickii (apical pore, three intercalary plates about equal in size, 1a pentagonal, 2a and 3a in contact with 4″), punctate ornamentation.

INTERNAL FEATURES: Eyespots not seen (Lemmermann 1900b); brownish plastids small, discoid (Taft and Taft 1971).

SIZE: 35–51 µm long by 29–42 µm wide (Taft and Taft 1971); Lemmermann (1900b) stated cells were 41–51 µm long by 32–42 µm wide.

ECOLOGICAL NOTES: Collected more often in cool seasons (Whitford and Schumacher 1969), spring in northern lakes (Hecky et al. 1986), from under the ice in Swedish lakes (Rengefors 1998) and in North Dakota (Phillips and Fawley 2002); shown to produce allelopathic substances (Rengefors and Legrand 2001).

Peridinium allorgei

Lefèvre 1927, 120–121 (no figures); 1932, 175, Figs. 852–855.

Slightly depressed round cell, sulcus wide to antapex. **PL26(1)**

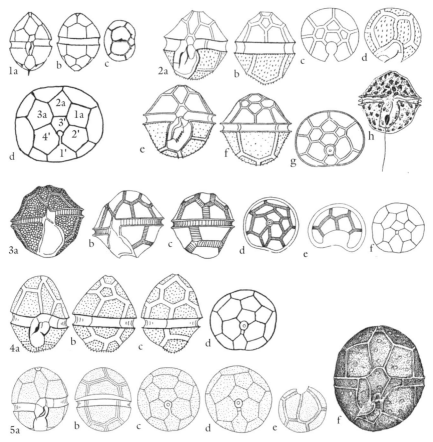

Plate 25. Species in the *Peridinium* Lomnickii group. 1. *P. aciculiferum*. (a) Ventral; note single hypotheca spine (b) Dorsal (c) Hypotheca (Lemmermann 1910, 663, Figs. 25–27) (d) Epithecal plate pattern. 2. *P. godlewski*. Note hypothecal spines. (a) Ventral (b) Dorsal (c) Epitheca (d) Hypotheca (Wołoszyńska 1916, Pl. 13, Figs. 31, 32, 35, 36) (e) Ventral with chloroplasts and eyespot (f) Ventral (g) Dorsal (h) Epitheca (Thompson 1950, Figs. 89–93). 3. *P. keyense*. (a, b) Ventral (c) Dorsal (d) Epitheca (e) Hypotheca (Nygaard 1926, Pl. 4, Fig. 32) (f) Epithecal plate pattern. 4. *Chimonodinium lomnickii*. Note punctate ornamentation of epitheca grading into small spines on hypotheca, lumpy antapex. (a) Ventral (b) Dorsal (c) Side (d) Epitheca (Wołoszyńska 1916, Pl. 10, Figs. 25–27, 29). 5. *P. wierzejskii*. Note round shape, punctate ornamentation. (a) Ventral (b) Dorsal (c, d) Epithecal variations (e) Hypotheca (Wołoszyńska 1916, Pl. 11, Figs. 1, 4, 5, 7, 8) (f) Ventral (Mary Ann Rood, with permission).

EXTERNAL FEATURES: Plate pattern: apical pore, 4′, 3a, 7″, 5‴, 2⁗; cell round to slightly depressed; apical plates asymmetrical, the 1a four sided; cingulum median without displacement; sulcus slightly penetrates the epitheca then broadly widens in the hypotheca, reaching the antapex; antapical plates unequal in size; only the 1‴ and 5‴ are visible in ventral view; ornamentation of longitudinal striae.

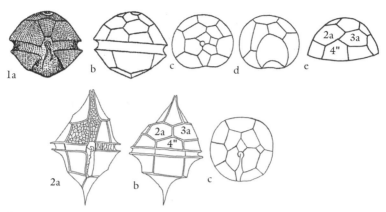

Plate 26. Species in the *Peridinium* Allorgei group. 1. *P. allorgei.* (a) Ventral (b) Dorsal (c) Epitheca (d) Hypotheca (Lefèvre 1932, Figs. 852–855) (e) Dorsal configuration (orig.). 2. *P. wisconsinense.* (a) Ventral (b) Dorsal (c) Epitheca (orig.).

SIZE: 25–40 µm long by 28–43 µm wide (Lefèvre 1927).

LOCATION REFERENCE: Not reported from North America.

Peridinium bipes

Stein 1883, Taf. 11, Figs. 7, 8.

Looks like *Peridinium willei* with an apical pore (wide lists at apex, antapex, and cingulum). **PL27(1); Fig. 112**

EXTERNAL FEATURES: Plate pattern: apical pore, 4′, 3a, 7″, 5‴, 2″″; oval cell with rounded hypotheca; dorsoventrally compressed; large 1′ plate reaching apex; narrow 2′ and 4′ plates; 1a, 3′, and 3a in a row; large 2a plate above 4″; lists on apical plates; antapical plates/sulcal margins with lists giving a posterior bilobed appearance; epitheca larger than hypotheca; cingulum offset about one cingulum width; sulcus deeply penetrating epitheca (at least one cingulum width) and expanding to antapex.

INTERNAL FEATURES: Chloroplasts many, small, parietal.

SIZE: 38–55 µm by 40–80 µm.

ECOLOGICAL NOTES: There have been reports of *P. bipes* causing red tides in Japan (Park and Hayashi 1993, Hirabayashi et al. 2007) and having an inhibitory effect on the cyanobacterium *Microcystis aeruginosa* (Wu et al. 1998).

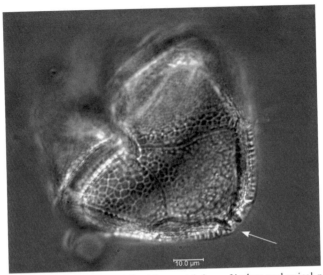

Figure 112. *Peridinium bipes.* Ventral epitheca with large 1′ plate and apical pore (*arrow*) (LM) (courtesy of John D. Hall).

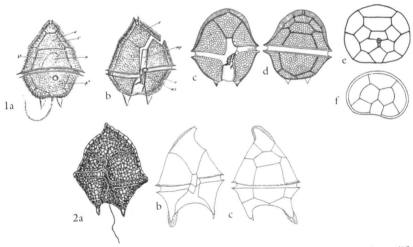

Plate 27. Species in the *Peridinium* Bipes group. 1. *P. bipes.* (a) Dorsal (b) Ventral (modified from Stein 1883, Taf 11, Figs. 7, 8) (c) Ventral; note large 1′ plate, apical pore (d) Dorsal, note apical lists and posterior spines (Tiffany and Britton 1952, Pl. 85, Figs. 988–991, with permission) (e) Epitheca (f) Hypotheca. 2. *P. limbatum.* (a) Ventral (Stokes 1887, Fig. 6) (b) Ventral (c) Dorsal (orig.).

Peridinium cinctum

(O.F.M.) Ehrenberg 1830, 38.

BASIONYM: *Vorticella cincta* O.F. Müller 1773, 105.
Plate pattern established by Stein 1883, Taf. 12,
Figs. 9–19.

Cell with thick thecal plates, without apical pore, apical and apical intercalary plates similar in size.
PL28(1a,b,c)

EXTERNAL FEATURES: Plate pattern: 4′, 3a, 7″, 5C,
5S, 5‴, 2‴′; round to oval; slight dorsoventral compression; 1′ plate not reaching apex; apical intercalary plates arranged so the 1a and 2a are smaller and to the left side, 3a is long and across the back; median cingulum slightly displaced (half to one cingulum width); sulcus entering epitheca and widening as it descends to antapex; there is a sulcal spine extending from the first postcingular plate.

INTERNAL FEATURES: Plastids golden brown, numerous, parietal; eyespot associated with chloroplast not noticeable in light microscope; nucleus in epitheca; accumulation bodies may be present (Calado et al. 1999).

SIZE: 50–73 µm long by 35–55 (up to 75) µm wide.

ECOLOGICAL NOTES: Planktonic. In Newfoundland it was found from the end of October until May, with maximum numbers under the ice in March (Davis 1973).

COMMENTS: Since *P. gatunense* has the same plate pattern and is more common, identifications as *P. cinctum* are questionable ("*P. cinctum*" in Thompson 1947 is *P. gatunense*). *Peridinium cinctum* is a taxonomically important species since it is the type species of one of the earliest dinoflagellate genera. Fensome et al. (1993) have gone as far as to say that taxa lacking the asymmetrical placement of apical plates may not belong in the genus *Peridinium* (pl28).

Peridinium cinctum f. *meandricum*

Lefèvre 1925. Rev. Algol. 2:334.

BASIONYM: *Peridinium meandricum* Brehm 1907 (description only).
Vermiform ornamentation on the hypotheca. **PL28(1g)**

LOCATION REFERENCE: TX (Carty 1986).

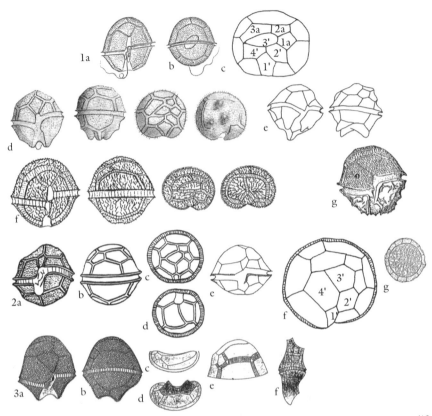

Plate 28. Species in the *Peridinium* Cinctum group. Species all with same plate pattern, differ in amount of dorsoventral compression, from none (*P. gatunense*) to moderate (*P. cinctum*) to great (*P. raciborskii*). 1. *P. cinctum*. (a) Ventral; note transverse flagellum indicated by cilia (b) Dorsal (modified from Stein 1883, Taf 12, Figs. 9, 10) (c) Plate pattern (d) *P. cinctum* f. *tuberosum*. Note three lobes in posterior. Ventral, dorsal, epitheca, hypotheca (Meunier 1919, Pl. 18, Figs. 23–26) (e) Right ventral, dorsal (Prescott 1951b, Pl. 91, Figs. 7, 11) (f) *P.cinctum* f. *westii*. Note vermiform ornamentation on plates. Ventral, dorsal, epitheca, hypotheca (Lemmermann in West and West 1905, text fig. p 495, Figs. 2 A–D) (g) *P. cinctum* f. *meandricum*. Note vermiform ornamentation only on hypotheca plates (Lefèvre 1932 Fig. 235). 2. *P. gatunense*. (a) Ventral (b) Dorsal (c) Epitheca; note circular cross section (d) Hypotheca (Ostenfeld and Nygaard 1925, Figs. 5, 6, 8, 10) (e) Ventral with small 1′ plate (f) Epitheca with plate annotations (orig.) (g) Cyst (Thompson 1947 as *P. cinctum*, Pl. 4, Fig. 5). 3. *P. raciborskii*. (a) Ventral (b) Dorsal (c) Apical from dorsal side (d) Hypotheca (e) Dorsal (f) Side view (Wołoszyńska 1912, Figs. 21 A–F).

Peridinium cinctum var. *tuberosum*

(Meunier) Lindemann 1928.

BASIONYM: *Peridinium tuberosum* Meunier 1919.
Hypotheca formed into three lobes. **PL28(1d,e)**

Peridinium cinctum f. *westii*

(Lemmermann) Lefèvre1925.

BASIONYM: *Peridinium westii* Lemmermann in West and West 1905, Figs. 2A–D, 495.

Large *P. cinctum* but differing in a large 2a plate, unequal antapical plates (2'''' larger than 1''''), with vermiform reticulation (Lefèvre 1925). Seen by Whitford and Schumacher (1969), who noticed the distinctive posterior ornamentation and thought the species larger (70 µm × 70 µm) than *P. cinctum*. **PL28(1f)**

LOCATION REFERENCE: IA (Prescott 1927), NC (Whitford and Schumacher 1969).

Peridinium gatunense

Nygaard in Ostenfeld and Nygaard 1925, 10–11, Figs. 5–10.

Saturn-shaped cell, slight anterior-posterior depression, wide cingular lists, small 1' plate. Common. **PL28(2); Figs. 113, 114; CP8(1,2)**

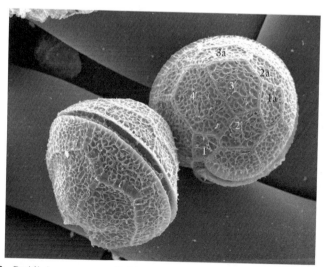

Figure 113. *Peridinium gatunense.* Epitheca and side views, the epitheca showing the small 1' plate and round outline (SEM).

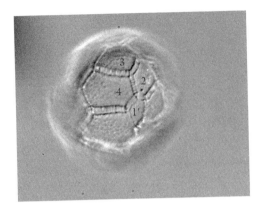

Figure 114. Empty cell, ventral epitheca with small 1′ plate evident (LM).

EXTERNAL FEATURES: Plate pattern: no apical pore, 4′, 3a, 7″, 5‴, 2⁗; *Peridinium gatunense* has the same plate pattern as *P. cinctum* but can be distinguished by overall cell appearance; cell large, wider than long because of depression and cingular lists; no apical pore; apical plates are asymmetrical, a small 1a and 2a and large dorsal 3a; 3a is larger than the squarish 4″ beneath it; 1′ plate small, the 3′ is the topmost plate, and large 2′ and 4′ plates meet in a ventral suture; sulcus barely enters the epitheca; reticulate ornamentation on thick plates.

INTERNAL FEATURES: Cells are golden to brown, no eyespot.

SIZE: 52–80 μm diameter, 36–80 μm long.

ECOLOGICAL NOTES: *P. gatunense* is widespread in the summer phytoplankton; the large, brownish, Saturn-shaped cells tumbling across the viewing field are distinctive. Reproduction has been studied (Pfiester 1977, Spector et al. 1981) and much research conducted on its ecology in Lake Kinneret, Israel (Hickel and Pollinger 1988, Berman-Frank and Erez 1996, Zohary et al. 1998).

Peridinium godlewskii

Wołoszyńska 1916, 274, Taf. 13, Figs. 31–36.

Hypothecal plates with short spiny ornamentation, plates concave. **PL25(2)**

EXTERNAL FEATURES: Plate pattern: apical pore, 4′, 3a, 7″, 5‴, 2⁗, cell ovoid, no dorsoventral compression; epitheca tapered to large apical pore, hypotheca rounded. Group Lomnickii (apical pore, three intercalary plates about equal in size, 1a pentagonal, 2a and 3a in contact with 4″), cingulum median to slightly above median (larger hypotheca), without displacement, sulcus entering epitheca less than one cingulum width and skewed to the right in the hypotheca without reaching the antapex, theca thin, plates concave, hypotheca with numerous fine spines on plates and sutures, first antapical plate larger than second.

INTERNAL FEATURES: Photosynthetic; plastids small, numerous, radial, with eyespot in sulcus.

SIZE: 25–35 µm long by 25–32 µm wide (Thompson 1950).

LOCATION REFERENCE: KS (Thompson 1950), NC (Whitford and Schumacher 1984).

Peridinium gutwinskii

Wołoszyńska 1912, 701–702, Fig. 22.

1a and 2a (sometimes also 3a) in contact with 4″ which is centered on dorsal surface. **Figs. 115, 116**

EXTERNAL FEATURES: Plate pattern: apical pore, pp, cp, 4′, 3a, 7″, 5C, 5S, 5‴, 2⁗; cell ovoid; apex notched at apical pore; 1a and 2a plates in contact with 4″ (this is the only species with this configuration); cingulum median with slight displacement; sulcus enters the epitheca about one cingulum width and does not reach the antapex; antapical plates equal; reticulate ornamentation.

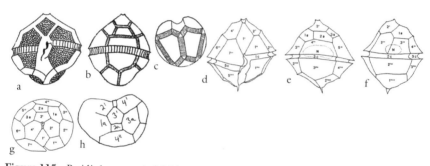

Figure 115. *Peridinium gutwinskii.* Note 1a and 2a in contact with 4″. (a) Ventral (b) Dorsal; note 3a also slightly touching the 4″ (c) Hypotheca (Wołoszyńska 1912, Fig. 22) (d) Ventral (e) Dorsal (f) Side (g) Epitheca with 3a distant from 4″ (Boltovskoy 1989, Figs. 1–4, used with permission of Borntraeger Cramer Science Publishers) (h) Interpretation of plates in Fig. 116.

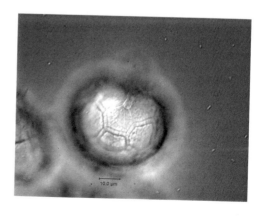

Figure 116. Epitheca with 1a, 2a, and 3a in contact with 4″ plate (courtesy of John D. Hall).

Wołoszyńska (1912) showed a smooth antapex, Boltovskoy (1989) showed a single antapical spine and sulcal plate configuration.

INTERNAL FEATURES: Numerous chloroplasts; nucleus in epitheca.

SIZE: 40–60 μm by 35–60 μm.

LOCATION REFERENCE: Panama (Prescott 1951a), PR (JH2009.01.25-2).

Peridinium keyense

Nygaard 1926, 208, Table 4, Fig. 32.

Large, round cell, plates smooth. **PL25(3)**

EXTERNAL FEATURES: Plate pattern: apical pore, 4′, 3a, 7″, 5‴, 2⁗; round cell; median cingulum, slightly displaced; sulcus penetrates epitheca about one cingulum width then widely expands in the hypotheca to the antapex; cingular spine on the right side and plates concave; antapical plates unequal, the first larger than the second. Group Lomnickii (apical pore, three intercalary plates about equal in size, 1a pentagonal, 2a and 3a in contact with 4″), intercalary plates 3a>2a>1a.

SIZE: 45–57 μm long by 47–54 μm wide.

LOCATION REFERENCE: Reported from NC as common in one lake (Whitford and Schumacher 1984).

Peridinium limbatum

(Stokes) Lemmermann 1900a.

BASIONYM: *Protoperidinium   limbatum*   Stokes 1888, Pl. 4, Fig. 1.

Two posterior horns, apical horn slightly bent. **PL27(2); Figs. 117, 118; CP8(3,4)**

EXTERNAL FEATURES: Plate pattern: apical pore, 4′, 3a, 7″, 5‴, 2⁗; conical epitheca; golden cells with large 1′ plate, extended into a short horn curved to the left; narrow 2′ and 4′ plates; 1a, 3′, and 3a in a row, large 2a plate above 4″; lists on plates give a fringe to the entire cell body; cingulum offset at least one cingulum width; sulcus does not penetrate the epitheca; two posterior horns composed of 1‴ and 1⁗ and 5‴ and 2⁗.

INTERNAL FEATURES: Chloroplasts yellowish brown.

SIZE: Cells 71–80 μm long by 61–67 μm wide.

ECOLOGICAL NOTES: Reports find it in soft (acid), brown-water lakes (Whitford and Schumacher 1969, Prescott 1951b), with *Sphagnum* (Stokes 1888, Prescott 1951b), and a bog pond (Pfiester and Skvarla 1980). It is found in the same habitats as *Ceratium carolinianum*. Yan and Stokes (1978) found it dominating the plankton in an acidic lake (ph 4.5–6) in

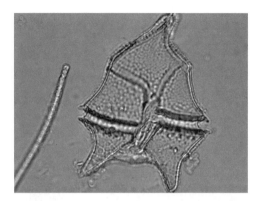

Figure 117. *Peridinium limbatum.*
Empty cells showing plates, slight tilt
of apex, and posterior horns, ventral.

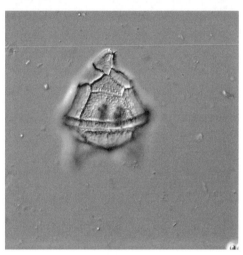

Figure 118. Dorsal (LM).

Ontario. Smith (1961) found it dominant, sometimes forming a film, in a
brown-water, dystrophic lake in Nova Scotia. Graham et al. (2004) considered it the dominant dinoflagellate in Crystal Bog (WI) in the summer, with
a doubling time of 10.7 days.

Peridinium lomnickii

Wołoszyńska 1916, 267–268, Taf. 10, Figs. 25–29.

Punctate ornamentation, fringe of antapical spines,
antapex lumpy. **PL25(4); Figs. 119–121**

EXTERNAL FEATURES: Plate pattern: apical pore, 4′,
3a, 7″, 5‴, 2⁗; Group Lomnickii (apical pore,
three intercalary plates about equal in size, 1a
pentagonal, 2a and 3a in contact with 4″); cell ovoid;
epitheca larger than hypotheca; cingulum without

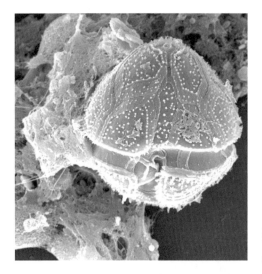

Figure 119. *Peridinium lomnickii.* Ventral epitheca with apical pore and punctate ornamentation.

Figure 120. Epithecal plate pattern.

Figure 121. Ventral hypotheca (SEM).

displacement; sulcus does not enter epitheca and descends in hypotheca with parallel sides not reaching antapex; plate ornamentation of irregular papillae that enlarge slightly in hypotheca; antapex asymmetrical and lumpy looking.

INTERNAL FEATURES: Chromatophores small, numerous; nucleus oval and central (Wołoszyńska 1916).

SIZE: 25–40 µm long by 22–35 µm wide.

TAXONOMIC NOTE: *Peridinium lomnickii* has been transferred to *Chimonodinium* based on differences from the type species of *Peridinium* (*P. cinctum*), partial LSU rDNA sequences, and internal features (Craveiro et al. 2011). Many species in *Peridinium* are not like the type species, and the dissolution of *Peridinum* is expected; however, there are other species in the Lomnickii group (as used by Lefèvre 1932, Popovský and Pfiester 1990, Starmach 1974), including *P. aciculiferum,* which is close to *lomnickii* in sequence but thought not to be a sister taxon (Craveiro et al. 2011), *P. keyense* and *P. godlewskii,* which were not discussed, and *P. wierzejskii,* which was reduced to a variety without examination of cells. *P. wierzejskii* and *C. lomnickii* are quite different in appearance, and reducing the former to a variety is premature.

Peridinium raciborskii

Wołoszyńska 1912, 700, Fig. 21.

SYNONYM: *Peridinium palustre* (Lindemann) Lefèvre 1932.

Dorsoventrally compressed *P. cinctum,* two antapical lobes. **PL28(3)**

EXTERNAL FEATURES: Plate pattern: no apical pore, 4′, 3a, 7″, 5‴, 2″″; oval cell; greatly dorsoventrally compressed (narrowly oval in top view); 1′, plate not to apex; 1a and 2a plates small and on the left side, 2a four sided, large 3a across the dorsal surface; cingulum submedian, displaced at least one cingulum width; sulcus may enter epitheca, widens in hypotheca to antapex; antapical plates form two lobes posteriorly.

INTERNAL FEATURES: Plastids numerous, radial (Popovský and Pfiester 1990).

SIZE: Cells 80–100 µm by 70–90 µm (Popovský and Pfiester 1990).

ECOLOGICAL NOTES: Collected from organically rich ponds and dystrophic lakes (Meyer and Brook 1969).

Peridinium striolatum

Playfair 1919, 810, Pl. 41, Fig. 1.

No apical pore, 1′ in contact with 2′, 3′, and 4′. **Fig. 122**

EXTERNAL FEATURES: Plate pattern: no apical pore, 4′, 3a, 7″, 5‴, 2⁗; cell oval; epitheca smoothly rounded, hypotheca may be flattened; small 1′ plate rounded, oval to ovate, in contact with 3′ plate; symmetrical arrangement of intercalary plates, 1a and 3a smaller, and the 2a across the dorsal surface, astride the 4″; cingulum slightly submedian; sulcus with deep penetration into the epitheca (at least two cingulum widths) extending to antapex, deeply concave, forming a furrow; a spine extends from the first postcingular plate over the sulcus; anatapical plates are equal in size; ornamentation of longitudinal striations.

SIZE: 41–57 μm long by 34–53 μm wide.

LOCATION REFERENCE: Not reported from North America.

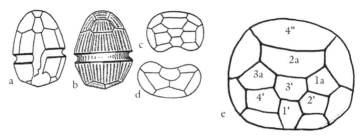

Figure 122. *Peridinium striolatum.* No apical pore, 1′ in contact with 3′. (a) Ventral (b) Dorsal with ornamentation (c) Apical (d) Antapical (Playfair 1919, Pl. 41, Fig. 1 a–d) (e) Epitheca (modified from Popovský and Pfiester 1990, Fig. 5).

Peridinium subsalsum

Ostenfeld 1908, 166–167, Pl. 5, Figs. 50–53.

1′ plate skewed to right; 4″ in contact with 2a, 3′, and 3a. **Fig. 123**

EXTERNAL FEATURES: Plate pattern: apical pore, 4′, 3a, 7″, 5‴, 2⁗; pentagonal cell; epitheca bluntly rounded, base flattened with small spines; cingulum below median, offset one cingulum width; sulcus only in hypotheca, widening from cingulum to antapex; anterior intercalary plates disjunct, 1a and 2a on the left side, 3a on the right; plate sutures not straight but curved; punctate ornamentation in vertical rows.

INTERNAL FEATURES: Nucleus central; chloroplasts radially arranged in epitheca; storage products in hypotheca (Popovský 1970).

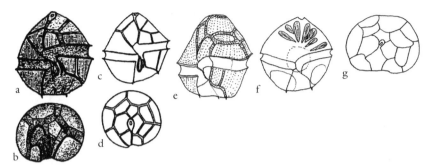

Figure 123. *Peridinium subsalsum.* Note apical pore, six-sided 4″. (a, c) Ventral (b) Hypotheca (d) Epitheca (Ostenfeld 1908, Pl. 5, Figs. 50–53) (e) Ventral with plates (f) Ventral with plastids and nucleus (g) Epitheca (Popovský 1970, Pl. 31, Figs. 4c, d, e).

SIZE: 22–40(60) µm long by 20–40(56) µm diameter.

LOCATION REFERENCE: Cuba (Popovský 1970).

Peridinium volzii

Lemmermann 1906, 166–167, Taf. 11, Figs. 15–18.

Symmetrical dorsal plates like *P. willei* but with smaller 1′ plate (does not reach apex); large Sa plate. **PL29(1a–d); Figs. 124, 125; CP8(5)**

EXTERNAL FEATURES: Plate pattern: no apical pore, 4′, 3a, 7″, 5‴, 2⁗; oval cell, dorsoventral compression with concave sulcus gives bean shape when viewed from above; symmetrical apical plates, with 2′ and 4′ about the same size and shape, 1a, 3′, and 3a in a row, large 2a plate aligned with 4″; 1′ plate five sided, flattened base at top of sulcus; deeply incised cingulum median with little displacement but strongly interrupted by sulcal plate Sa in epitheca; sulcus penetrates epitheca one cingulum width with large Sa plate, has parallel sides through the hypotheca to the antapex; antapical plates about equal in size; reticulate ornamentation.

INTERNAL FEATURES: Golden-brown plastids discoid, parietal; no eyespot.

SIZE: 42–60 µm long by 38–60 µm diameter.

ECOLOGICAL NOTES: Popovský and Pfiester (1990) synonymized *P. volzii* with *P. willei,* but I consider them separate species. For more information on biology see Berdach 1977 (cells identified as *Peridinium cinctum* but are *Peridinium volzii*), Pfiester and Skvarla 1979, Hayhome et al. 1987.

Figure 124. *Peridinium volzii.*
Ventral.

Figure 125. Epitheca with 2', 3',
and 4' similar size, 1a and 3a similar
and 2a long (SEM).

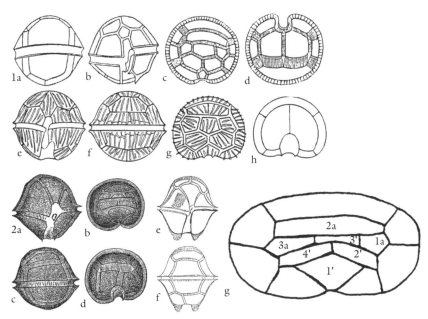

Plate 29. Species in the *Peridinium* Willei group. 1. *P. volzii.* (a) Dorsal (b) Ventral; note smaller 1' (c) Epitheca (d) Hypotheca (Lemmermann 1906, Pl. 11, Figs. 14–17) (e) *P. volzii* f. *vancouverense.* Ventral (f) Dorsal (g) Epitheca (h) Hypotheca (Wailes 1928a as *Peridinium striolatum*, Pl. 2, Figs. 4–7). 2. *P. willei.* (a) Ventral; note large 1' plate (b) Epitheca with symmetrical arrangement of plates (c) Dorsal (d) Hypotheca (Huitfeldt-Kaas 1900, Figs. 6–9) (e) Ventral, highlighting the apical and antapical lists (f) Dorsal (Thompson 1947, Pl. 3, Figs. 5–6) (g) Plate pattern.

Peridinium volzii f. *vancouverense*

(Wailes) Lefèvre 1932.

BASIONYM: *Peridinium vancouverense* Wailes 1934.

SYNONYM: *Peridinium striolatum* in Wailes 1928a, Pl. 2, Figs. 4–7.

Cells distinguished by distinctive ornamentation of parallel striae on each plate. **PL29(1e–h)**

SIZE: Large, up to 60 µm diameter (Bourrelly 1966).

ECOLOGICAL NOTES: Rare in bogs (Duthie et al. 1976).

LOCATION REFERENCE: AB (Thomasson 1962), Labrador (Duthie et al. 1976), ON (Bourrelly 1966).

Peridinium wierzejskii

Wołoszyńska 1916, 269, Taf. 11, Figs. 1–8.

Round cell, no dorsoventral compression. **PL25(5); Figs. 126, 127**

EXTERNAL FEATURES: Plate pattern: apical pore, cp, 4′, 3a, 7″, C6, 5‴, 2‴‴; Group Lomnickii (apical pore, three intercalary plates about equal in size, 1a pentagonal, 2a and 3a in contact with 4″ or 2a and 3a in contact with 5″); 1′ plate does not reach apex; punctate ornamentation; cingulum median without much displacement; sulcus extending into epitheca about one cingulum width as large Sa plate, Sd forming flap, not reaching antapex.

INTERNAL FEATURES: Disk-shaped chloroplasts, nucleus central (Wołoszyńska 1916).

SIZE: 37–40 μm long, 32–35 μm diameter (G. LaLiberte pers. comm.).

ECOLOGICAL NOTES: Abundant in lakes in northern Wisconsin (G. LaLiberte pers. comm). In cells examined from both Michigan and Ecuador, the cingulum was not incised, may have been a cyst.

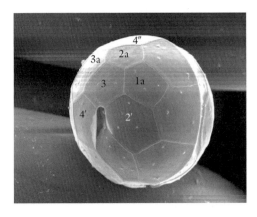

Figure 126. *Peridinium wierzejskii.* Round cell with Lomnickii pattern, little ornamentation (SEM).

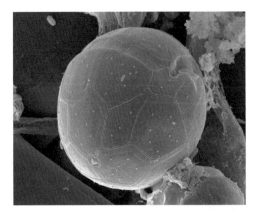

Figure 127. Ventral apical view with apical pore, sulcus (LM).

LOCATION REFERENCE: FL (SC 031208 L. Michelle), MI (SC08-19-03, Watersmeet), WI (Gina LaLiberte)

TAXONOMIC NOTE: A revision of *Peridinium lomnickii* to *Chimonodinium lomnickii* (Craveiro et al. 2011) included reducing *Peridinium wierzejskii* to a variety of *C. lomnickii* without examination of cells. *Peridinium wierzejskii* is a round cell, whereas *C. lomnickii* is more oval with a bump at the antapex, and they differ in ornamentation. I consider the removal of *Peridinium wierzejskii* from *Peridinium* and placement in *Chimonodinium* premature.

Peridinium willei

Huitfeldt-Kaas 1900, 5, Figs. 6–9.

Large 1' plate extends to apex; no apical pore; prominent apical, antapical, and cingular lists; symmetrical arrangement of apical plates; dorsoventrally compressed. Common. **PL29(2); Figs. 128–130; CP8(6)**

EXTERNAL FEATURES: Plate pattern: no apical pore, 4', 3a, 7", 5"', 2""; large; this commonly found planktonic *Peridinium* (the other being *P. gatunense*) is immediately recognized by its lists. Cell is dorsoventrally compressed with apical plates in a symmetrical arrangement; 1' plate large, reaching the cell apex; 2' and 4' form a narrow strip; 1a, 3', and 3a form a parallel strip. The 2a is large and matches sutures with the large rectangular 4" beneath it; apical plates (upper margin of 1', apical margins of 2' and 4', 1a, 3', 3a) all with lists. There are cingular lists, and posterior plates have lists that give the cell a bilobed appearance; cingulum is median and displaced

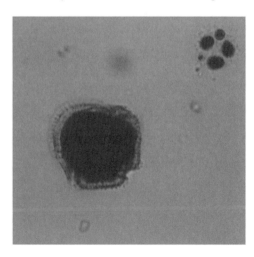

Figure 128. *Peridinium willei.*
Stained cell with apical lists and bilobed posterior evident (LM).

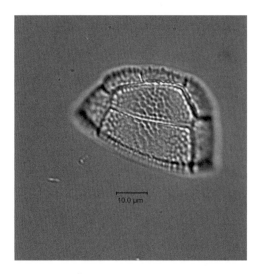

Figure 129. Dorsal epitheca of empty cell showing large 4′ and 2a plates.

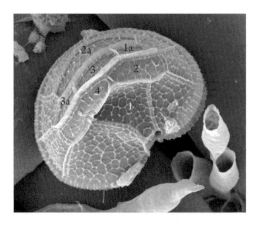

Figure 130. Epitheca showing large 1′ to apex, narrow apical and apical intercalary plates (SEM).

1–2 cingulum widths; sulcus only slightly interrupted by cingulum, then descends with parallel sides to the antapex.

INTERNAL FEATURES: Plastids golden to brown, numerous, parietal; no eyespot.

SIZE: Cell 49–68 μm diameter, 50–70 μm long.

ECOLOGICAL NOTES: Studies on Danish lakes found *P. willei* at a broad range of pH (4.2–8.5), year-round, with a maximum in late summer (Olrik 1992).

Peridinium wisconsinense

Eddy 1930, 300–301, Fig. 51.

Spindle-shaped cell. **PL26(2); Figs. 131–134; CP8(7–9)**

EXTERNAL FEATURES: Plate pattern: apical pore, 4', 3a, 7", S4, C5+T, 5''', 2''''; no dorsoventral compression; chunky central portion tapering to an apical horn and antapical horn; 1' plate extending from sulcus to apical pore, 2', 3', and 4' plates about

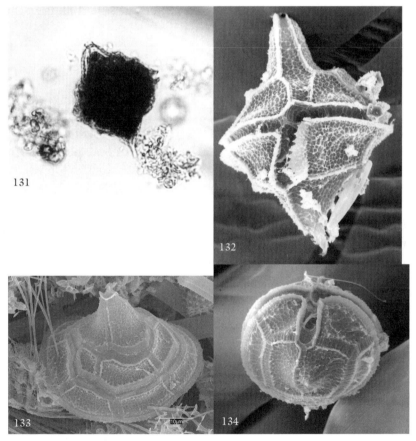

Figure 131. *Peridinium wisconsinense.* Stained cell showing rounded apex and pointed antapex (LM).

Figure 132. Ventral view (SEM).

Figure 133. Dorsal epitheca (SEM).

Figure 134. Hypotheca (SEM).

equal in size, arranged around pore; three anterior intercalary plates variable in size and shape. Eddy (1930) illustrated 1a small, 2a medium, 3a large; Carty (1986) shows a small, four-sided 1a; 2a and 3a larger and about equal in size. Sulcus not entering epitheca, not reaching antapex, wide sulcal list on the left side; cingulum not displaced; reticulate ornamentation on the plates. The appearance of the apical pore is unique, long ridge extends along the ventral margin of the 4' plate to the apex, another ridge encircles the pore including the terminus of the long ridge; hypothecal horn from 1''''.

INTERNAL FEATURES: Golden chloroplasts, color may be masked by red droplets (Nicholls 1973).

SIZE: 55–64 μm long by 48–56 μm wide.

ECOLOGICAL NOTES: Planktonic, may be abundant (Whitford and Schumacher 1969), may be found under the ice (Phillips and Fawley 2002).

Prorocentrum

Ehrenberg 1834.

TYPE SPECIES: *Prorocentrum micans* Ehrenberg 1834.

Theca of two large plates; both flagella apical; no obvious cingulum or sulcus.

Prorocentrum is primarily a marine genus although there are some freshwater species (Croome and Tyler 1987). One species reported from North America, *Prorocentrum compressum*.

Prorocentrum compressum

(Stein) Carty comb. nov.

BASIONYM: *Dinopyxis compressa* Stein 1883, Taf. 1, Figs. 34–38.

SYNONYM: *Exuviaella compressa* (Stein) Ostenfeld 1903.

Theca of two large plates; both flagella apical; no obvious cingulum or sulcus.
Fig. 135

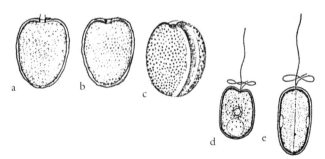

Figure 135. *Prorocentrum.* (a, b) Front/back view (c) Side view showing valves (Stein 1883, Taf 1, Figs. 34–36) (d) Front view with apical flagella (e) Side view (Thompson 1950, Figs. 10, 11 as *Exuviella compressa*).

EXTERNAL FEATURES: Thecate, oval, laterally flattened, photosynthetic cell with two large plates (valves), no cingulum or sulcus, apex of cell with two flagellar pores surrounded by small plates.

INTERNAL FEATURES: Two green to yellow parietal plastids, each with a pyrenoid, nucleus posterior, no eyespot.

SIZE: 22–26 µm long by 15–18 µm wide, 11–12 µm thick (Thompson 1950).

ECOLOGICAL NOTES: Collected from a swamp.

LOCATION REFERENCE: MD (Thompson 1950).

Sphaerodinium

Wołoszyńska 1916, 279–280.

TYPE SPECIES: *Sphaerodinium polonicum* Wołoszyńska 1916, 280, Taf. 14, Figs. 1–22.

Plate pattern: apical pore, 4′, 4a, 7″, 6‴, 2″″; with an eyespot, photosynthetic.

Sphaerodinium is considered by some (Loeblich III 1980, Fensome et al. 1993) a junior synonym of *Glenodinium* Ehrenberg, based on a desire to maintain the genus *Glenodinium* even though it was described without plates and based primarily on a horseshoe-shaped eyespot. *Sphaerodinium* has been accepted by the major compendia (Starmach 1974, Popovský and Pfiester 1990). Its unique plate pattern, different from that first proposed for *Glenodinium cinctum*, sets it apart from *Glenodinium*. Large-subunit rDNA

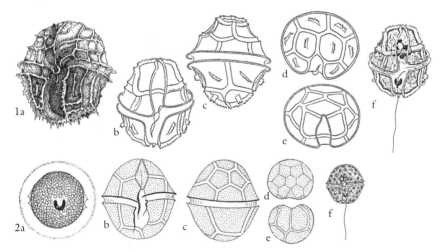

Plate 30. Species of *Sphaerodinium*. 1. *S. fimbriatum*. Note lists extending from plate margins and in the center of hypotheca plates. (a) Ventral (Carty 2003, Fig. 6Ca) (b) Ventral (c) Dorsal (d) Hypotheca (e) Epitheca (f) Ventral (Thompson 1950, Figs. 63–67). 2. *S. polonicum*. (a) Ventral view with horseshoe-shaped eyespot (b) Ventral (c) Dorsal (d) Epitheca; note regular arrangement of apical plates (e) Hypotheca (Wołoszyńska 1916, Taf 12, Figs. 1, 2, 3, 5, 6) (f) Ventral with chloroplasts, eyespot, and flagella (Thompson 1950, Fig. 58).

analysis of a species of *Sphaerodinium* placed it remote from *Peridinium* and closer to woloszynskoid taxa (Craveiro et al. 2010).

Key to species of *Sphaerodinium*

1a. Cell round, smooth .... *S. polonicum*
1b. Cell with fimbriate extension of plate margins and in center of hypothecal plates...
.... *S. fimbriatum*

Sphaerodinium fimbriatum

Thompson 1950, 296–297, Figs. 63–67.

Fimbriate extensions of plate margins and hypothecal midplate fimbriate extensions. Rough edges of this species distinguish it in the light microscope. **PL30(1); Figs. 136–139**

EXTERNAL FEATURES: Plate pattern: apical pore, 4′, 4a, 7″, C7, S6, 6‴, 2⁗; oval cell, dorsoventral compression; slender, elongate 1′ plate flanked by

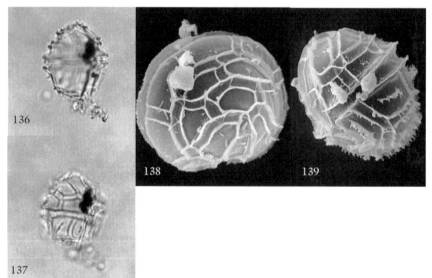

Figure 136. *Sphaerodinium fimbriatum*. Empty cell; note rough outline due to extended plate margins and fimbriae.

Figure 137. Empty cell with plates visible and fimbriae on hypotheca plates (LM).

Figure 138. Epitheca with wide sutures between plates (SEM).

Figure 139. Side view with fimbriae on postcingular plates and plate margins extended (SEM) (Carty 2003, Fig. 7A).

similarly shaped 2′ and 4′ plates on the ventral face, 3′ alone visible on the dorsal face; deeply incised cingulum slightly above median, displaced one cingulum width, sulcus in hypotheca only, with parallel sides to antapex, sutures double with raised rims, apical pore area triangular. Hypotheca remarkable, double suture rims extended into fringe, fringed lists on all posterior plates, margins of sulcus fringed. Thompson described only hypotheca plates with a "median fimbriate costa" but they may be found to a lesser extent on epithecal plates as well.

INTERNAL FEATURES: There are golden plastids and a horseshoe-shaped eyespot; red accumulation bodies in the epitheca.

SIZE: 42–53 μm long by 42–46 μm wide (Thompson 1950), 33–41 μm long by 32–33 μm wide (TX, Carty 1986).

ECOLOGICAL NOTES: Collected during the summer, may be abundant in some samples.

Sphaerodinium polonicum

Wołoszyńska 1916, 280, Taf. 14, Figs. 1–22.

Spherical cell. **PL30(2); Fig. 140**

EXTERNAL FEATURES: Plate pattern: 4′, 4a, 7″, 6‴, 2⁗; spherical cell with slight to no dorsoventral compression; 3′ plate six sided; 2′, 4′ and all four apical intercalary plates pentagonal and of similar size; median cingulum with slight displacement; sulcus penetrates epitheca about one cingulum width and does not reach the antapex; sulcal margin

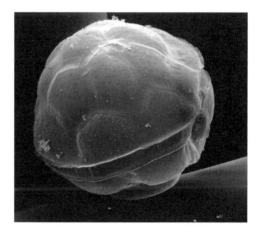

Figure 140. *Sphaerodinium polonicum*. Round cell with light punctate ornamentation (SEM).

of the 1‴ plate bears a spine; smooth to punctate ornamentation. Note: an apical pore is not clearly present in the drawings of Thompson (1950) nor the micrograph of Carty (Carty and Wujek 2003) but is slightly illustrated by Wołoszyńska (1916).

INTERNAL FEATURES: Plastids golden brown, numerous, parietal; large eyespot in sulcus. (Thompson 1950).

SIZE: 38–45 μm long by 37–42 μm wide. Cell 25 μm wide and deep (Belize, Carty and Wujek 2003), diameter 43–45 μm (IL, Tiffany and Britton 1952).

Staszicella

Wołoszyńska 1916, 277.

TYPE SPECIES: *Staszicella dinobryonis* Wołoszyńska 1916.

Associated with *Dinobryon;* epitheca smaller than hypotheca.

Freshwater, thecate dinoflagellate associated with *Dinobryon*. Plate pattern: apical pore, 4′, 1a, 7″, 5‴, 2⁗, at least three cingular plates, epitheca smaller than hypotheca.

Staszicella dinobryonis

Wołoszyńska 1916, 278, Taf. 12, Figs. 32–40.

SYNONYMS: *Glenodinium dinobryonis* (Wołoszyńska) Schiller 1937; *Peridiniopsis dinobryonis* (Wołoszyńska) Bourrelly 1968a.

Associated with *Dinobryon* colonies. **Fig. 141**

EXTERNAL FEATURES: Plate pattern: apical pore and canal plate, 5′, 1a, 7″, 5‴, 2⁗; apical plates asymmetrical (Lindemann 1925 shows great variability

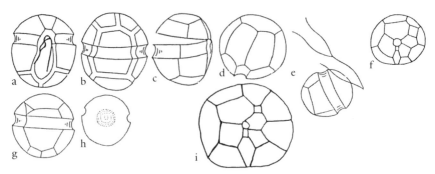

Figure 141. *Staszicella dinobryonis.* (a) Ventral (b) Dorsal (c) Thecal separation (d) Hypotheca (e) Cell associated with *Dinobryon* cell (f) Epitheca (g) Dorsal (h) Location of nucleus (Wołoszyńska 1916, Taf 12, Figs. 34–40) (i) Epithecal pattern (in manner of Popovský and Pfiester 1990).

in apical plate sizes, number, and arrangements); wide cingulum divides oval cell into smaller epitheca and larger hypotheca; antapical plates large.

INTERNAL FEATURES: Nucleus round, central; chloroplasts present; eyespot probably present (Wołoszyńska 1916).

SIZE: 22–35 µm by 19–38 µm.

ECOLOGICAL NOTES: Cells attached to lorica of *Dinobryon divergens* (Bourrelly 1966).

LOCATION REFERENCE: QC (Bourrelly 1966), UT (Rushforth and Squires 1985).

Thompsodinium

Bourrelly 1970, 81.

TYPE SPECIES: *Thompsodinium intermedium* (Thompson) Bourrelly 1970, 81.

Cell with huge postcingular plates 2‴, 3‴, and 4‴ that extend to the antapex. Plate pattern: apical pore, 4′, 3a, 6″, C6, S4, 5‴, 2⁗, eyespot present. Postcingular plates 2‴, 3‴, and 4‴ large, reaching antapex, freshwater, photosynthetic.

Thompsodinium intermedium

(Thompson) Bourrelly 1970

BASIONYM: *Peridinium intermedium* Thompson 1950, 298, Figs. 80–88.

SYNONYM: *Bagredinium crenulatum* Coutè and Iltis in Da et al. 2004.

Cell with huge postcingular plates 2‴, 3‴, and 4‴ that extend to the antapex, antapical list. **Figs. 142–147; CP7(7,8)**

EXTERNAL FEATURES: Plate pattern: apical pore, 4′, 3a, 6″, C6, S4, 5‴, 2⁗; ovate cell; slight dorsoventral compression; epitheca tapering to apical pore, hypotheca rounded; first apical plate spindle-shaped, 2′, 4′, and all three apical intercalaries pentagonal; cingulum median without displacement; sulcus penetrates epitheca one cingulum width and expands widely in hypotheca almost to antapex; sulcal plate Sa enters the epitheca, Sd is a flap over the flagellar pore, small Ss, large Sp not to antapex. Relationship between 3′ and 2a plate most often *conjunctum* (seen in samples from Cuba, Belize, TX, MI, NY) but *remotum* and *contactum* also seen. Hypothecal plates distinctive, the 1‴ is reduced due to large Sp plate; 2‴, 3‴, and 4‴ are huge plates that reach the antapex; both antapical plates are small and border

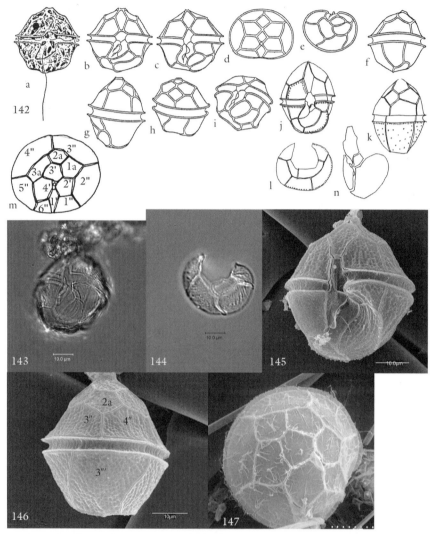

Figure 142. *Thompsodinium intermedium.* Note huge postcingular plates. (a) Ventral with plastids, eyespot, and flagella (b, c) Ventral plates (d) Epitheca (e) Hypotheca (f, g, h, i) Different views of postcingular plates (Thompson 1950, Figs. 80–88) (j) Ventral (k) Dorsal (l) Hypotheca (m) Epitheca with plates indicated (n) Sulcal plates (original).

Figure 143. *Thompsodinium intermedium.* Empty cell, ventral hypotheca (LM).

Figure 144. Hypotheca with small antapical plate and large postcingular plates (LM).

Figure 145. Ventral view with distinctive 1″″ plate almost a part of the sulcus, a hint of the fringe of posterior teeth that can sometimes be seen in light micrographs.

Figure 146. Dorsal view with huge 3‴ plate.

Figure 147. Epitheca with apical and apical intercalary plates in a star pattern.

the Sp; suture of 1″″ and 2‴ with list; plates with light reticulate to punctate or vermiform ornamentation.

INTERNAL FEATURES: Plastids pale to deep golden brown.

SIZE: 32–39 µm long by 33–40 µm wide.

ECOLOGICAL NOTES: Formed a bloom (TX, Gilpin 2012) in a river during low-flow summer conditions, found almost unialgal in tiny pool near Caribbean (Belize).

Tovellia

Moestrup, Lindberg, and Daugbjerg in Lindberg et al. 2005, 427.

TYPE SPECIES: *Tovellia coronata* (Wołoszyńska) Moestrup, Lindberg, and Daugbjerg in Lindberg et al. 2005.

Roundish photosynthetic cell with eyespot, with numerous, thin, hexagonal plates.

This genus was extracted from *Woloszynskia* based on *W. coronata*, for species with numerous, thin, usually hexagonal plates; an apical line of narrow plates (ALP); an eyespot not enclosed in a membrane; and a lobed cyst with terminal spines (Lindberg et al. 2005).

Tovellia coronata

(Wołoszyńska) Moestrup, Lindberg, and Daugbjerg in Lindberg et al. 2005, 427.

BASIONYM: *Gymnodinium coronatum* Wołoszyńska 1917, 115–116, Taf. 11, Figs. 10–19, Taf. 13, Figs. I–L, N

SYNONYM: *Woloszynskia coronata* (Wołoszyńska) Thompson 1950, 290–291.

Many large, thin, hexagonal plates; red coloration.

EXTERNAL FEATURES: Roundish cell with relatively large, irregularly positioned plates; cingulum median, slightly displaced; sulcus does not enter epitheca and is a shallow groove not reaching antapex; double row of narrow rectangular plates begins in the middle of the apical, ventral face and continues over the apex to the dorsal side. Wołoszyńska (1917) illustrated a single, large, hexagonal, antapical plate with ornamentation.

INTERNAL FEATURES: Golden to brown chloroplasts; red eyespot; work with cells in culture found them to contain red globules in the cytoplasm and a nucleus in the hypotheca (Lindberg et al. 2005).

SIZE: 19–39 µm long by 14–36 µm wide;

Tovellia glabra

(Wołoszyńska) Moestrup, Lindberg, and Daugbjerg in Lindberg et al. 2005, 429.

BASIONYM: *Gymnodinium coronatum* var. *glabrum* Wołoszyńska 1917, Taf. 11, Figs. 20, 21.

Round, golden cell with many small plates. **PL12(3); Figs. 148–150**

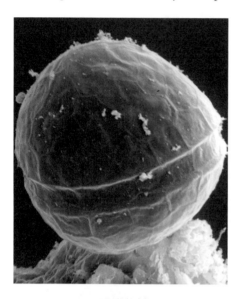

Figure 148. *Tovellia glabra*. Side view with small polygonal plates barely visible (Carty 2003, Fig. 8J).

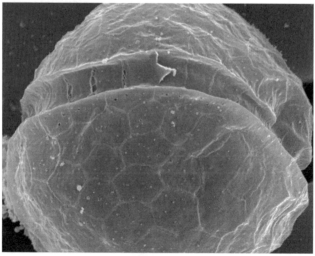

Figure 149. Polygonal plates cover cell.

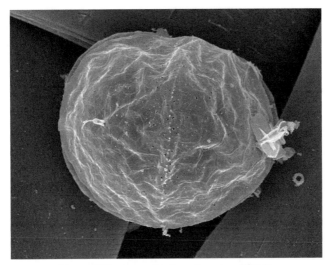

Figure 150. Apex with line of punctae (SEM).

EXTERNAL FEATURES: This species differs from *T. coronata* by the presences of several antapical plates instead of a single plate. Wołoszyńska (1917) provided no details about other differences; see species description of *T. coronata*. Lindberg et al. (2005) raised the variety to species level. Examination of reports of *Woloszynskia coronatum* by Thompson (1950) and Carty (1986), reveals cells lacking the single antapical plate, thus *T. glabra.* Most of my records of this genus come from SEM micrographs, the antapical plate is often not evident; I am calling all records *T. glabra* since *T. coronata* is described as being red (Lindberg et al. 2005) and these were golden.

Tyrannodinium

Calado, Craveiro, Daugbjerg, and Moestrup 2009, 1202–1203.

TYPE SPECIES: *Tyrannodinium berolinense* (Lemmermann) Calado, Craveiro, Daugbjerg, and Moestrup 2009, 1203.

Heterotrophic, no eyespot.

Plate formula: apical pore, canal plate, 4', 0a, 6", 6C, 5+S, 5''', 2''''; plates thin; cingulum median with little displacement; large nucleus in hypotheca; heterotrophic feeding on algae, nematodes, and the like, via a feeding tube (peduncle) extending from sulcus; food vacuoles in epitheca. Freshwater genus placed in estuarine family Pfiesteriaceae based on thecal morphology and mode of feeding (Calado et al. 2009).

Tyrannodinium berolinense

(Lemmermann) Calado, Craveiro, Daugbjerg, and Moestrup 2009, 1203.

BASIONYM: *Peridinium berolinense* Lemmermann 1900c, 308–309 (description only).

SYNONYMS: *Glenodinium berolinense* (Lemmermann) Lindemann 1925; *Peridiniopsis berolinense* (Lemmermann) Bourrelly 1968a, 9.

Heterotrophic, small antapical spine(s). **PL15(6)**

EXTERNAL FEATURES: Plates thin, difficult to see; plate pattern: apical pore slightly off center, canal plate, 4′, 0a, 6″, 5‴, 2⁗; cell angularly biconical; little to no dorsoventral compression; cingulum about median (hypotheca larger than epitheca in older cells), without displacement; sulcus not penetrating epitheca, with parallel sides, not reaching antapex; small antapical spine(s) on suture between plates; apical notch; apical plates not symmetrical, large 4′ and medium 2′ compress the small 3′; the five-sided 4″ is in a central dorsal position so plates seem skewed in dorsal or apical views; sulcal list associated with antapical plate.

INTERNAL FEATURES: No eyespot, nonphotosynthetic though may have pigmented accumulation bodies, cytosol granular, large nucleus in hypotheca.

SIZE: 30–37 µm long by 29–32 µm wide (Thompson 1950).

ECOLOGICAL NOTES: Species recognized in part by behavior, cells congregate around algae, protozoa, or injured larger prey (Calado et al. 2009).

LOCATION REFERENCE: KS (Thompson 1950), WI (Wedemayer and Wilcox 1984).

COMMENTS: Calado (2011) synonymized *Tyrannodinium berolinense* with *Peridiniopsis edax*, and since the specific epithet *edax* has priority, he formed *Tyrannodinium edax*. Examination of plate patterns by Thompson (1950), who saw both species and focused on differences, shows two different tabulation patterns and only *T. berolinense* with an apical pore, so I am maintaining both species.

Woloszynskia

Thompson 1950, 286–288.

TYPE SPECIES: *Woloszynskia reticulata* Thompson 1950 (designated by Loeblich and Loeblich 1966).

Theca of numerous thin polygonal plates; photosynthetic.

Cells mostly roundish, golden; thecal plates not obvious. Most of the species were transferred from *Gymnodinium* (athecate genus) based on the presence of a thin theca when the cell ecdyses. Recent work has shown that *Woloszynskia* is polyphyletic based on eyespot construction, cyst type, and molecular sequences, and species have been moved into other genera (Lindberg et al. 2005, Moestrup et al. 2008); see "Taxonomy" in the introduction. The type species remains, as do species not specifically moved to other genera.

Key to species of *Woloszynskia*

1a. Epithecal plates thin, hypothecal thick, with eyespot .... *W. reticulata*
1b. Epithecal and hypothecal plates similar 2
2a. Reddish pigment globules present, winter species 3
2b. No reddish pigment globules .... *W. cestocoetes*
3a. Cell small 12–15 µm long, four plastids .... *W. ordinata*
3b. Cell larger, many plastids 4
4a. Cell posterior truncate to bilobed .... *W. pascheri*
4b. Cell posterior rounded .... *W. vera*

Woloszynskia cestocoetes

(Thompson) Thompson 1950, 291.

BASIONYM: *Gymnodinium cestocoetes* Thompson 1947, 8–9, Pl. 1, Figs. 10–14.

Cell obovoid; sulcus to antapex; cryophile. **PL31(1)**

EXTERNAL FEATURES: Obovate cell, slight dorsoventral compression, hemispherical epitheca, tapering hypotheca to flattened antapex, deep cingulum slightly above median, overhung by epitheca, without displacement, sulcus in hypocone, narrow below cingulum, widening to antapex. Reproduction occurs when motile cell settles, divides; the daughter cells become motile before complete separation (Thompson 1947). Thompson (1950) moved this species he named as *Gymnodinium* to *Woloszynskia* without comment. Original drawings and description do not mention plates (though plates are illustrated for another *Gymnodinium* species mentioned in the same article, *Gymnodinium neglectum*). I think it was the mode of reproduction that convinced Thompson that this was a species of *Woloszynskia*.

INTERNAL FEATURES: No eyespot; numerous, small, peripheral, radially arranged plastids; red oil globules.

SIZE: 23–25 µm long by 19–21 µm wide.

ECOLOGICAL NOTES: Collected January through March from under the ice in a swamp.

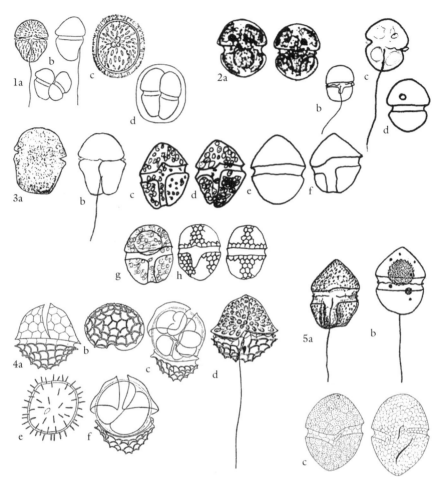

Plate 31. Species of *Woloszynskia.* 1. *W. cestocoetes.* (a) Ventral view (b) Side (c) Cyst (d) Motile cell dividing (Thompson 1947, Pl. 1, Figs. 10–14 as *Gymnodinium cestocoetes*). 2. *W. ordinata.* (a) Ventral views with large chloroplasts in each quadrant (Skuja 1939, Taf 10, Figs. 26, 27 as *G. ordinatum*) (b) Ventral view showing sulcus not reaching antapex (Forest 1954a, Fig. 511, as *G. ordinatum*) (c) Ventral view (d) Dorsal (Thompson 1947, Pl. 1, Figs. 3, 4 as *G. ordinatum*). 3. *W. pasheri.* (a) Cell outline (b) Ventral view with slightly bilobed posterior (Suchlandt 1916, text figs. B, D as *G. pasheri*) (c, d) Ventral views with slightly different amounts of sulcal penetration into epicone and numerous chloroplasts (e) Dorsal (f) Ventral without cell contents (g) Ventral (h) Cell covering of numerous plates (redrawn from Schiller 1954, Figs. 24a–d, 28a, 29a,b). 4. *W. reticulata.* (a) Dorsal view with apical slit (b) Hypothecal view with thick plate margins (c) Ecdysing to release gametes (d) Ventral view with peripheral, radially arranged chloroplasts (e) Spiny cyst (f) Ecdysing to release two daughter cells (Thompson 1950, Figs. 15–20). 5. *W. vera.* (a) Ventral (b) Dorsal (Lindemann 1925 as *Gymnodionium veris,* Figs. 71, 72) (c) Two ventral views showing plates (Wołoszyńska 1917, Pl. 11, Figs. 1, 2 as *Gymnodinium carinatum* var. *hiemalis*).

LOCATION REFERENCE: MD (Thompson 1947), WI (Graham et al. 2004).

Woloszynskia ordinata

(Skuja) Thompson 1950, 291.

BASIONYM: *Gymnodinium ordinatum* Skuja 1939, 151, Taf. 10, Figs. 26–28.

Small cell, four plastids. **PL31(2)**

EXTERNAL FEATURES: Cell ovate; epicone and hypocone broadly rounded; dorsoventrally compressed; cingulum median, without displacement; sulcus not entering epicone and only slightly into hypocone; longitudinal flagellum up to twice cell length.

INTERNAL FEATURES: Four parietal chromatophores, yellow to olive brown; red oil globules; no eyespot.

SIZE: 12–15 μm long by 10–13 μm wide, 8 μm thick (Thompson 1947, Forest 1954a).

ECOLOGICAL NOTES: Collected in February from under ice (MD, Thompson 1947), from a tundra pond (Sheath and Hellebust 1978). Many reports from EPA lists.

Woloszynskia pascheri

(Suchlandt) Stoch 1973, 126, 133.

BASIONYM: *Glenodinium pascheri* Suchlandt 1916, Figs. A–F, 244.

SYNONYMS: *Gymnodinium pascheri* (Suchlandt) Schiller 1954.

Causes red snow. **PL31(3)**

EXTERNAL FEATURES: Ovoid cell with rounded epicone; squarish hypocone; cingulum slightly above median, with slight displacement; sulcus not entering epicone, slightly penetrating hypocone.

INTERNAL FEATURES: Red pigment globules; nucleus in epicone; greenish brown plastids peripheral.

SIZE: 20–38 μm long by 13–24 μm wide, 12–18 μm thick.

ECOLOGICAL NOTES: Collected from red snow along ditch near marshy area, especially near dogwoods and in ditch when temperature of water was below 10 °C (Gerrath and Nicholls 1974).

LOCATION REFERENCE: ON (Gerrath and Nicholls 1974).

Woloszynskia reticulata

Thompson 1950, 288, Figs. 15–20.

Thick ridges delineate hexagonal plates in the hypotheca. **PL31(4); Figs. 151–153; CP7(9)**

EXTERNAL FEATURES: Conical epitheca, rounded hypotheca; dorsoventral compression; apical carina/slit bisecting epitheca; cingulum median with slight displacement; epitheca with four or five series of plates with delicate sutures; hypotheca with four (19–20 postcingular, 14 postcingular intercalary, 9 antapical intercalary, 3–4 antapical) series of plates with thick sutures; sulcus only in hypotheca, not to antapex.

INTERNAL FEATURES: Chloroplasts yellow brown; stigma in the sulcus. Cyst spherical, spiny in appearance (spines are plates attached along one side and perpendicular to original position) (Thompson 1950).

SIZE: Cells 25–52 μm long by 21–46 μm wide.

ECOLOGICAL NOTES: Swims by a twirling, rocking motion; collected during summer months (June–September).

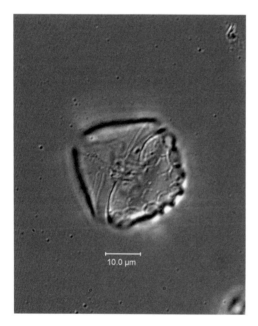

10.0 μm

Figure 151. *Woloszynskia reticulata.* Ventral view; hypotheca strongly scalloped because of thick plate margins, conical epitheca (LM).

Figure 152. Epitheca of numerous polygonal plates; note flagellum in cingulum (*arrow*).

Figure 153. Ventral view showing difference between plate margin thickness in the epitheca and that in the hypotheca.

Woloszynskia vera

(Lindemann) Thompson 1950, 290.

BASIONYM: *Gymnodinium veris* Lindemann 1925, 155, Figs. 71, 72.

SYNONYM: *Gymnodinium carinatum* Schill. var. *hiemalis* Wołoszyńska 1917, Taf. 11, Figs. 1, 2, Taf. 12, Fig. 12.

Cell obovoid, sulcus not obvious in hypocone. **PL31(5)**

EXTERNAL FEATURES: Cell obovoid, cingulum median; sulcus not obvious; theca composed of numerous small plates; hypotheca strongly dorsoventally compressed (Wołoszyńska 1917).

INTERNAL FEATURES: Eyespot small (Lindemann 1925); nucleus central (Wołoszyńska 1917).

SIZE: 18–35 µm long by 28 µm wide, cells sometimes much smaller (Wołoszyńska 1917).

ECOLOGICAL NOTES: Found in sewage-polluted Arctic lake (Kalff et al. 1975).

LOCATION REFERENCE: NU (Kalff et al. 1975 as *Gymnodinium veris*).

Appendix A
Documentation of Taxa by Location

The citations for each species are grouped by area in the following order: Greenland, Canada, United States (first citations, then personal communications), Mexico, Central America, the Caribbean.

Actiniscus canadensis
NT (Bursa 1969)

Amphidiniopsis sibbaldii
ON (Nicholls 1998)

Amphidinium klebsii
MA (Herdman 1924), MD (Thompson 1950), NJ (Barlow and Triemer 1988)

Amphidinium luteum
NWT (Sheath and Steinman 1982, Sheath and Hellebust 1978), ON (Kling and Holmgren 1972
WA (Larson et al. 1998)

Amphidinium wigrense
WI (Wilcox and Wedemayer 1985)

Bernardinium bernardinense
MN (Meyer and Brook 1969), TX (SC 830107-3)

Borghiella dodgei
Greenland (Moestrup et al. 2008)
KY (Dillard 1974)

Ceratium brachyceros

NC (as *Ceratium hirundinella* f. *brachyceras*, Whitford and Schumacher 1969), OH (Carty 1993), SC (Goldstein and Manzi 1976 as *Ceratium hirundinella* f. *brachyceras*), TX (Carty 1986)

Oaxaca (Figueroa-Torres and Moreno-Ruiz 2003)

Ceratium carolinianum

BC (Wailes 1934), NL (Duthie et al. 1976), Newfoundland (Davis 1972), NS (Hughes 1947–1948, as *C. curvirostre*), ON (Duthie and Socha 1976), QC (Brunel and Poulin 1992)

FL, GA (Bailey 1850), MA (Prescott and Croasdale 1937, Wall and Evitt 1975), MN (Meyer and Brook 1969), NC (Whitford and Schumacher 1969), SC (Bailey 1850), WI (Prescott 1951b), TX (Carty 1986)

MD (JH2006.06.23-3), MI (SC/VF Watersmeet, 030819), NY (JH2009.07.25-9).

Ceratium cornutum

BC (Wailes 1934), NS (Hughes 1947–1948), ON (Kling and Holmgren 1972)

AR (Meyer 1969), IL (Tiffany and Britton 1952), MI, WI (Prescott 1951b).

Ceratium furcoides

AR (Hilgert et al. 1978a), CA (M Morris et al. 1979b), GA (F Morris et al. 1978), IA (Williams et al. 1979c), KS (Williams et al.1979b), LA (Lambou et al. 1979c), MT (Hern et al. 1979b), MO (M Morris et al. 1979a), NC (M Morris et al. 1978b), ND (Taylor et al. 1979b), NE (F Morris et al. 1979b), NM (Lambou et al. 1979a), NV (Lambou et al. 1979b), OK (Hern et al. 1979a), SD (Hern et al. 1979c), UT (Williams et al. 1979d), WA (F Morris et al. 1979a)

Michoacan (Ortega 1984 as *Ceratium hirundinella* f. *furcoides*)

Cuba (Popovský 1970 as *Ceratium hirundinella* f. *furcoides*)

NY (SC Greenport 8-16-97).

Ceratium hirundinella

AB (Thomasson 1962), BC (Wailes 1934), NL (Duthie et al. 1976), MB (Wailes 1934), NB, NS (Hughes 1947–1948), NT (Sheath and Steinman 1982), ON (Duthie and Socha 1976), PEI (Hughes 1947–1948), QC (Brunel et Poulin 1992), SK (Hammer et al. 1983), YT (Thomasson 1962)

AK (Wailes 1934, Hooper 1947), AL (Taylor et al. 1977), AR (Meyer 1969), CA (Wall and Evitt 1975), CO (Whitford and Kim 1971), CT (Huszar et al. 2003), FL (Van Meter-Kasanof 1973), GA (Schumacher 1956), IA (Prescott 1927), ID (F Morris et al. 1979c), IL (Tiffany and Britton 1952), IN (Wailes 1934), KY (Dillard and Crider 1970), MA (Auyang 1962), MD (Thompson 1947), MI (Prescott 1951b), MN (Meyer and Brook 1969), MS (Williams et al. 1977), MT (Hilgert 1976), NC (Whitford and Schumacher 1969), NH (Gruendling and Mathieson 1969), NJ (Williams et al. 1978), NM (Lambou et al. 1979a), NV (Lambou et al. 1979b), NY (Huszar et al. 2003), OH (Taft and Taft 1971, Carty 1993), OR (Thomasson 1962), OK (Vinyard 1958), PA (Herbst and Hartman

1981), RI (Hanisak 1973), SC (Goldstein and Manzi 1976), TN (Forest 1954a), TX (Carty 1986), UT (Rushforth and Squires 1985), VA (Bovee 1960), WI (Prescott 1951b), WA (Schumacher and Muenscher 1952), WV (Woodson 1969), WY (Thomasson 1962)

DF, Hidalgo (Ortega 1984), Guanajuato (Lopez-Lopez and Serna-Hernandez 1999), Jalisco, Estado de Mexico (Ortega 1984), Michoacán (Tafall 1941), Morelos, Oaxaca, Puebla, Tamaulipas, Veracruz (Figueroa-Torres and Moreno-Ruiz 2003)

Guatemala (Clark 1908, Ostenfeld and Nygaard 1925), Panama (Ostenfeld and Nygaard 1925)

Cuba (Laiz et al. 1993b)

Ceratium hirundinella f. *austriacum*
AB (Thomasson 1962)

IL (Eddy 1930), MT (Thomasson 1962, Hilgert 1976), OR (Thomasson 1962), SD (Hern et al. 1979c), WY (Thomasson 1962).

Ceratium hirundinella f. *brachyceroides*
OH (SC 970716-12 Medusa Marsh)

Michoacán (Tafall 1941)

Ceratium hirundinella f. *carinthiaceum*
AB (Thomasson 1962), BC (Wailes 1934)

IA (Prescott 1927), IN (Wailes 1934), MT, OR (Thomasson 1962), WA (F Morris et al. 1979a).

Ceratium hirundinella f. *piburgense*
BC (Wailes 1934)

CA (Thomasson 1962), IN (Wailes 1934), MN (Meyer and Brook 1969), NC (Whitford and Schumacher 1969), OH (Carty 1993), OR (Taylor et al. 1979c), SC (Goldstein and Manzi 1976), WA (Wailes 1934), WI (Eddy 1930)

Ceratium hirundinella f. *robustum*
BC (Wailes 1934)

AR (Hilgert et al. 1978a), IA (Prescott 1927), IL (Eddy 1930), NH (Gruendling and Mathieson 1969), ND (Taylor et al. 1979b), WY (Williams et al. 1979a)

Michoacán (Tafall 1941)

Ceratium hirundinella f. *scotticum*
CO (M Morris et al. 1979c), ID (F Morris et al. 1979c), IL (M Morris et al. 1978a), KS (Williams et al. 1979b), MT (Hilgert 1976), NC (M Morris et al. 1978b), ND (Taylor et al. 1979b), NE (F Morris et al. 1979b), OH (Carty 1993), OR (Taylor et al. 1979c), SD (Hern et al. 1979c), UT (Williams et al. 1979d), WA (F Morris et al. 1979a)

Ceratium rhomvoides
  BC (Wailes 1934 as *Ceratium hirundinella* f. *gracile*)
  OH (Carty 1993 as *Ceratium hirundinella* f. *gracile*), TN (Johansen et al. 2007)
  CT (SC08), FL (SC950131 L Michelle), MA (SC 08)
  Michoacán (Tafall 1941 as *Ceratium hirundinella* f. *gracile*)

Cystodinedria inermis
  OH (Carty 1993), OK (Pfiester and Popovský 1979)
  NC (SC 090605-2, Charlotte), TN (SC060508), TX (SC 840407)

Cystodinium acerosum
  MD (Thompson and Meyer 1984)

Cystodinium bataviense
  KS, MD (Thompson 1949), MN (Meyer and Brook 1969), OH (Taft and Taft 1971), OK (Pfiester and Lynch 1980), TN (Forest 1954a)
  NC (Parrow, pers. comm.), TX (SC 820616C), VA (JH 2006.06.24-3)

Cystodinium iners
  BC (Stein and Borden 1979), NL (Duthie et al. 1976 as *C. cornifax*), ON (Duthie and Socha 1976)
  IL (Wunderlin 1971 as *Cystodinium cornifax*), KY (Dillard et al. 1976), KS, MA, MD (Thompson 1949), MI (Prescott 1951b), MN (Meyer and Brook 1969), MT (Hoham 1966), NC (Whitford and Schumacher 1969 as *Cystodinium cornifax*), OH (Taft and Taft 1971), SC (Goldstein and Manzi 1976 as *Cystodinium cornifax*), WI (Prescott 1951b as *Cystodinium cornifax*)
  NY (JH2008.06.07-5), TX (SC 830616-4)
  PR (Foerster 1971 as *Cystodinium cornifax*)

Cystodinium phaseolus
  MN (Meyer and Brook 1969)

Cystodinium steinii
  BC (Stein and Borden 1979), ON (Duthie and Socha 1976)
  MI (Prescott 1951b), MN (Meyer and Brook 1969), SC (Jacobs 1971), WI (Prescott 1951b)

Dinamoeba coloradense
  CO (Bursa 1970)

Dinastridium sextangulare
  TN (Forest 1954a)
  MD (SC 060623), TX (SC Brazos Co)

Dinococcus bicornis

MN (Meyer and Brook 1969 as *Raciborskia bicornis*), TN (Johansen et al. 2007), WI (Prescott 1951b as *Raciborskia bicornis*)

Panama (Prescott 1955)

Dinosphaera palustris

BC (Stein and Borden 1979), NT (Sheath and Steinman 1982)

FL (Van Meter-Kasanof 1973), IA (Prescott 1927), IL (Eddy 1930, Tiffany and Britton 1952), IN (Daily and Miner 1953), MA (Prescott and Croasdale 1937), MN (Meyer and Brook 1969), MT (Morgan 1971 as *Glenodinium palustre*), NC (Whitford and Schumacher 1969 as *Glenodinium palustre*), NH (Gruendling and Mathieson 1969 as *Gonyaulax palustris*), NY (Schumacher 1961), OK (Pfiester and Terry 1978), TN (Forest 1954a), VA (Forest 1954b), WI (Prescott 1951b)

Michoacán (Tafall 1941 as *Goniaulax palustre*, Ortega 1984)

Panama (Prescott 1951a as *Glenodinium palustre*)

Durinskia dybowski

NT (Sheath and Steinman 1982 as *Glenodinium dybowski*), NU, QC (Whelden 1947 as *Glenodinium dybowski*)

MN (Eddy 1930 as *Glenodinium dybowski*), TN (Johansen et al. 2007 as *Durinskia baltica*), TX (Carty 1986 as *Durinskia baltica*)

OH (SC/VF Fremont 920630)

Mexico City (year round, Rosaluz Tavera 2/2012)

Entzia acuta

MB (Hecky et al. 1986)

AZ (Taylor et al. 1979a), CA (M Morris et al. 1979b), KS (Williams et al. 1979b), OH (Taft and Taft 1971 as *Diplosalis acuta*), OK (Pfiester and Terry 1978), SD (Hern et al. 1979c), UT (Williams et al. 1979d)

MA (SC 080730-4), NE (SC/VF L. Johnson 8-4-03) OH (SC/VF Bellview 8-5-97)

Esoptrodinium gemma

KS (Thompson 1950), MD (Delwitch, pers. comm.), NC (Parrow, M Charlotte, Raleigh 2008), TN (SC 040713-6 Great Smokey Mts NP)

Glenodiniopsis steinii

BC (Dickman 1968 as *Gymnodinium ulignosa*)

MN (Highfill and Pfiester 1992a), VA (Bovee 1960 as *Gymnodinium ulignosum*)

NY (JH2009.07.18-2)

Glochidinium penardiforme

NB (SC 080804-2)

AR (Hilgert et al. 1978a), CA (M Morris et al. 1979b). KS (Thompson 1950), LA (Lambou et al. 1979c), MO (M Morris et al. 1979a), MS (Williams et al. 1977), NC (Whitford and Schumacher 1969), ND (Taylor et al. 1979b), UT (Rushforth and Squires 1985), WI (Prescott 1951b)

NE (SC/VF Jeffrey Canyon SWMA 8-4-03), OH (SC/VF Bellview 8-4-97), TX (SC 6-29-82)

Gloeodinium montanum

BC (Dickman 1968), NL (Duthie et al. 1976)

AK (Carty 2007), AR (Kelley and Pfiester 1989), MD (Thompson 1949), MN (Meyer and Brook 1969), MT (Hoham 1966), OH (Taft and Taft 1971)

CA (JH 2008.08.18-4), NY (JH 2006.06.09), PA (Michaux State Forest JH2008.05.17-1), TX (SC 840130-2)

PR (JH2009.01.30-8)

Gonyaulax apiculata

BC (Stein and Borden 1979 as *Gonyaulax* sp.)

NC (Whitford and Schumacher 1969)

Gonyaulax spinifera

NY (VF Sagaponack 8-16-97)

Gymnodinium acidotum

NWT (Sheath and Steinman 1982)

LA (Farmer and Roberts 1990), MO (Fields and Rhodes 1991), OH (Carty 1993, Klarer et al. 2000)

Gymnodinium aeruginosum

BC (Wailes 1934), ON (Duthie and Socha 1976), QC (Puytorac et al. 1972). NB (SC080803-4), NS (SC 080805-1)

MD (Thompson 1950), MN (Meyer and Brook 1969), NC (M Morris et al. 1978b), OH (Taft and Taft 1971), TN (Johansen et al. 2007), WI (Wilcox and Wedemayer 1984 as *Gymnodinium acidotum*)

MA (SC080730-2), ME (SC080802-4), NH (SC080801-2), NY (JH2006.06.09 Harriman SP), RI (SC 080807-6), TX (SC roadside ditch 820705-4), WA (Robin Matthews 8-08 Fazon Lake)

Gymnodinium albulum

AR (Hilgert et al. 1978a), IA (Williams et al. 1979c), ID (F Morris et al. 1979c), KS (Williams et al. 1979b) LA (Lambou et al. 1979c), MD (Thompson 1950), MO (M Morris et al. 1979a), MT (Hern et al. 1979b), ND (Taylor et al. 1979b), NE (F Morris et al. 1979b), OK (Hern et al. 1979a), SD (Hern et al. 1979c), UT (Williams et al. 1979d), WA (F Morris et al. 1979a)

MI (SC 030819-2), TX (SC 820622D)

Gymnodinium bogoriense
Costa Rica (Umaña Villalobos 1988)

Gymnodinium caudatum
AK (SC 060713-2 Juneau), WI (Prescott 1951b)

Gymnodinium cryophyllum
WI (Wedemayer et al. 1982)

Gymnodinium excavatum
NL (Duthie et al. 1976)

Gymnodinium fuscum
BC (Stein and Borden 1979), NL (Duthie et al. 1976), ON (Kling and Holmgren 1972), QC (Brunel and Poulin 1992), SK (Hammer et al. 1983). NB (SC080803-3), NS (SC 080805-2)

AK (Carty 2007), FL (Dawes and Jewett-Smith 1985), KY (Kobraei and White 1996), MD (Thompson 1947), MN (Meyer and Brook 1969), NH (Gruendling and Mathieson 1969), NC (Whitford and Schumacher 1969), OR (Taylor et al. 1979c), PA (Murphy et al. 1994), TN (Hill 1980), VA (Bovee 1960), WI (Prescott 1951b)

CT (SC 080728-4), ME (SC 080801-4), NY (JH2009.06.06.09), RI (SC 080807-11), TX (SC Jones Forest 831016-1), VT (SC 080731-8)

Gymnodinium helveticum
BC (Stein and Borden 1979), MB (Hecky et al. 1986), NU (Kalff et al. 1975), NWT (Sheath and Steinman 1982), ON (Schindler and Holmgren 1971, Findlay 2003)

MI (Carrick 2005), OH (Klarer et al. 2000)

TX (SC 830911)

Gymnodinium inversum
NWT (Sheath and Steinman 1982)

AK (Prescott and Vinyard 1965), MA (Wright 1964 as *Gymnodinium incurvum*), MN (Meyer and Brook 1969), MT (Hilgert 1976), OR (McIntire et al. 2007)

MI (SC 030819-2)

Gymnodinium lacustre
NU (Kalff et al. 1975), NWT (Sheath and Steinman 1982), ON (Duthie and Socha 1976)

TN (Johansen et al. 2007)

Gymnodinium limneticum
AL (Lackey 1936), IL (Wunderlin 1971), NC (Whitford and Schumacher 1969), TN (Forest 1954a), VA (Woodson and Seaburg 1983)

Gymnodinium luteofaba
   QC (Janus and Duthie 1979)

Gymnodinium marylandicum
   MD (Thompson 1947)

Gymnodinium mirable
   NU (Kalff et al. 1975), NWT (Sheath and Steinman 1982), ON (Kling and Holmgren 1972)

Gymnodinium palustre
   ON (Schindler and Holmgren 1971)
   AK (Alexander et al. 1980), IA (Prescott 1927), MA (Croasdale 1948), MI (Prescott 1951b), MN (Meyer and Brook 1969), MT (Prescott and Dillard 1979), NH (Gruendling and Mathieson 1969), NC (Whitford and Schumacher 1969), OH (Klarer et al. 2000), TN (Forest 1954a), VA (Woodson and Seaburg 1983)
   Baja California (Figueroa-Torres and Moreno-Ruiz 2003), Michoacán (Tafall 1941), Oaxaca (Figueroa-Torres and Moreno-Ruiz 2003)

Gymnodinium rotundatum
   QC (Brunel and Poulin 1992)
   MA (Wright 1964)

Gymnodinium thompsonii
   KS (Thompson 1950), VA (Marshall and Burchardt 2004)
   OH (SC)

Gymnodinium triceratium
   ON (Kling and Holmgren 1972)
   AK (Alexander et al. 1980), MD (Thompson 1947), NC (Whitford and Schumacher 1984), VA (Parson and Parker 1989),
   CA (JH 2008.08.18-1), NY (JH 2009.04.27-4), PA (JH 2007.11.12-1), TX (SC L. Sommerville 820622D)

Gymnodinium uberrimum
   BC (Dickman 1968), NWT (Sheath and Steinman 1982), ON (Kling and Holmgren 1972)
   AK (Alexander et al. 1980), MI (Sicko-Goad and Walker 1979)

Gymnodinium varians
   SK (Hammer et al. 1983)

Gymnodinium viride
BC (Wailes 1934, Stein 1975)
MD (Thompson 1947)

Gyrodinium pusillum
ON (Duthie and Socha 1976 as *Gymnodinium pusillum*)
MD (Thompson 1947)

Haidadinum ichthyphilum
BC (Buckland-Nicks et al. 1997)

Hemidinium nastum
BC (Stein and Borden 1979), NL (Duthie et al. 1976), QC (Puytorac et al. 1972), SK (Hammer et al. 1983)
AK (Alexander et al. 1980), AR (Meyer 1969), CA (Allen 1920), IA (Prescott 1927), KY (Kobraei and White 1996), MD (Thompson 1947), MN (Meyer and Brook 1969), NC (Whitford and Schumacher 1969), NH (Yeo 1971), OH (Taft and Taft 1971), OK (Pfiester and Terry 1978), TN (Forest 1954a), TX (Carty 1986), VA (Forest 1954b), WI (Prescott 1951b)
NY (JH2009.0)
Mexico Tabasco (Figueroa-Torres and Moreno-Ruiz 2003)

Hemidinium ochraceum
MD (Thompson 1947)

Hypnodinium sphaericum
QC (Brunel and Poulin 1992)
MD (Thompson 1947), MN (Meyer and Brook 1969), NC (Whitford and Schumacher 1969), OH (Taft and Taft 1971)
NY (JH2009.06.06)

Hypnodinium monodisparatum
MT (Hilgert 1976)

Jadwigia applanata
NU (Whelden 1947 as *Glenodinium neglectum*), QC (Puytorac et al. 1972 as *Gymnodinium neglectum*)
MD (Thompson 1947), KS (Thompson 1950), NC (Whitford and Schumacher 1969 as *Gymnodinium neglectum*), TN (Forest 1954a as *Gymnodinium neglectum* but mentions plates), VA (Bovee 1960, as *Glenodinium neglectum*)
OR (YP Waldo Lake), WA (culture FW145 from UWCC as *Woloszynskia limnetica* referenced in Lindberg et al. 2005)

Kansodinium ambiguum
   KS (Thompson 1950), TX (Carty 1986)

Katodinium auratum
   CO (Bursa 1970)

Katodinium bohemicum
   KS (Thompson 1950), TN (Johansen et al. 2007)

Katodinium fungiforme
   OH (Klarer et al. 2000)

Katodinium glandulum
   BC (Stein and Borden 1979)

Katodinium musei
   AR (Meyer 1969), MN (Meyer and Brook 1969)

Katodinium planum
   BC (Stein and Borden 1979)
   MD (Thompson 1947 as *K. musei*), TN (Johansen et al. 2007)

Katodinium spirodinioides
   OH (VF 970808-1)

Katodinium tetragonops
   KS (Thompson 1950)

Katodinium vorticella
   BC (Stein and Borden 1979)
   AK (Wailes 1934), IA (Prescott 1927 as *Gymnodinium vorticella*), KS (Thompson 1950), TN (Johansen et al. 2007)

Lophodinium polylophum
   TX (Carty 1986)
   Veracruz (Osorio-Tafall 1942)
   FL (C Delwiche/S Handy Ft Pierce 9/07)

Naiadinium biscutelliforme
   ON (Duthie and Socha 1976)
   AR (Hilgert et al. 1978a), AZ (Taylor et al. 1979a), CA (M Morris et al. 1979b), IA (Williams et al. 1979c), KS (*Thompson 1950), LA (Lambou et al. 1979c), MN (Meyer and Brook 1969 as *Glenodinium gymnodinium*), MO (M Morris et al. 1979a), MS (Canion and Ochs 2005), MT (Hern et al. 1979b), NE (F Morris

et al. 1979b), NH (Yeo 1971 as *Glenodinium gymnodinium*), NM (Lambou et al. 1979a), OK (Nolen et al. 1989), SD (Hern et al. 1979c), TN (Johansen et al. 2007), TX (*Carty 1986), WY (Williams et al. 1979a)

AK (SC060712-10 Juneau), MA (SC080730-4), MD (SC060624-5), ME (SC080801-7), VT (SC080731-5)

Michoacán (*Tafall 1941), Oaxaca (Figueroa-Torres and Moreno-Ruiz 2003)

* Includes illustration with 2a configuration.

Naiadinium polonicum

CA (M Morris et al. 1979b), IA (Williams et al. 1979c), ID (F Morris et al. 1979c), IL (Eddy 1930, M Morris et al. 1978a), LA (Lambou et al. 1979c as *Glenodinium gymnodinium*), MO (M Morris et al. 1979a), ND (Taylor et al. 1979b), OR (Taylor et al. 1979c), TN, WI (Eddy 1930), WY (Williams et al. 1979a)

Cuba (Popovský 1970)

Palatinus apiculatus

ON (Duthie and Socha 1976 as *Peridinium palatinum*)

IA (Prescott 1927 as *Peridinium marsonii*)

Michoacán (Tafall 1941)

Palatinus apiculatus var laevis

ON (Duthie and Socha 1976 as *Peridinium palatinum* var. *laeve*)

KS (Thompson 1950)

Palatinus pseudolaeve

NL (Duthie et al. 1976), ON (Duthie and Socha 1976 as *Peridinium pseudolaeve*)

Parvodinium africanum

ON (Kling and Holmgren 1972), QC (Puytorac et al. 1972)

MN (Meyer and Brook 1969), TX (Carty 1986)

FL (SC 3A3 7-29-94)

Parvodinium beliziense

Belize (Carty and Wujek 2003)

Parvodinium centenniale

BC (Wailes 1934)

Belize (Carty and Wujek 2003)

Parvodinium deflandrei

AB (Thomasson 1962)

MS (Canion and Ochs 2005), NY (CSLAP 2003), TN (Johansen et al. 2007)

Oaxaca (Figueroa-Torres and Moreno-Ruiz 2003)

Parvodinium goslaviense
NT (Sheath and Steinman 1982), ON (Schindler and Holmgren 1971), QC (Janus and Duthie 1979)

NH (Yeo 1971)

OH (SC/VF Delta Reservoir 970823)

Parvodinium inconspicuum
Greenland (Hansen 1967)

BC (Wailes 1934), NL (Duthie et al. 1976), MB (Shamess et al. 1985), NT (Sheath and Steinman 1982), ON (Schindler and Holmgren 1971, Findlay 2003), QC (Janus and Duthie 1979)

AK (Wailes 1934), AR (Hilgert et al. 1978a), AZ (Taylor et al. 1979a), CA (M Morris et al. 1979b), GA (F Morris et al. 1978), IA (Williams et al. 1979c), IL (Eddy 1930, Tiffany and Britton 1952), KY (Kobraei and White 1996), LA (Lambou et al. 1979c), MD (Thompson 1947), MI (Carrick 2005), MO (M Morris et al. 1979a), MT (Morgan 1971), MS (Canion and Ochs 2005), NC (Whitford and Schumacher 1969), ND (Taylor et al. 1979b), NE (F Morris et al. 1979b), NH (Gruendling and Mathieson 1969), OH (Carty 1993), OK (Pfiester et al. 1984), OR (McIntire et al. 2007), SC (Goldstein and Manzi 1976), TN (Johansen et al. 2007), TX (Carty 1986), UT (Williams et al. 1979d), VA (Parson and Parker 1989), WA (Larson et al. 1998), WI (Prescott 1951b), WV (Perez et al. 1994)

FL (SC L Michelle), MS (AC 2004), NY (JH2008.09.21-1)

Michoacán (Tafall 1941), Tamaulipas (Figueroa-Torres and Moreno-Ruiz 2003)

Cuba (Popovský 1970), Guadeloupe (Bourrelly and Manguin 1952)

PR (JH2009.02.01-6)

Costa Rica (Hargraves and Viquez 1981, Haberyan et al. 1995), Panama (Prescott 1951a)

Parvodinium lubieniense
NL (Duthie et al. 1976)

Parvodinium pusillum
BC (Wailes 1934), ON (Findlay 2003), QC (Brunel et Poulin 1992), SK (Hammer et al. 1983)

AR (Meyer 1969), CO (Kolesar et al. 2002), GA (F Morris et al. 1978), IA (Prescott 1927), IL (Tiffany and Britton 1952), KY (Kobraei and White 1996), LA (Lambou et al. 1979c), MI (Prescott 1951b), MN (Meyer and Brook 1969), NC (Whitford and Schumacher 1969), NH (Yeo 1971), SC (Goldstein and Manzi 1976), TN (Forest 1954a), VA (Woodson and Seaburg 1983), WI (Eddy 1930)

Cuba (Margalef 1947, Popovský 1970)

Parvodinium umbonatum
BC (Wailes 1934), NL (Duthie et al. 1976), ON (Duthie and Socha 1976), QC (Puytorac et al. 1972)

AR (Hilgert et al. 1978a), CO (Toetz and Windell 1993), FL (Taylor et al. 1978), GA (F Morris et al. 1978), ID (F Morris et al. 1979c), IL (Tiffany and Britton 1952), LA (Lambou et al. 1979c), MD (Thompson 1947), MN (Meyer and Brook 1969), OH (Carty 1993), OR (Taylor et al. 1979c), SD (Hern et al. 1979c), TN (Johansen et al. 2004), TX (Carty 1986)

AK (SC060710-1, Juneau), CA (JH2008.08.18-3), NM (Kalina Manoylov), NY (JH2008.07.18-7),PA(JH2008.05.24-1),VA(JH2005.10.16-1),WA(R.Williams), WV (JH2006.10.22-1)

Cuba (Popovský 1970), Guadeloupe (Bourrelly and Manguin 1952), PR (JH2009.01.25-3)

Estado de Mexico, Oaxaca (Figueroa-Torres and Moreno-Ruiz 2003), Yucatan (Schmitter-Soto et al. 2002)

Peridiniopsis borgei
ON (Duthie and Socha 1976), SK (Hammer et al. 1983 as *Peridinium borgei*)

MT (Morgan 1971 as *Glenodinium borgei*), NC (Whitford and Schumacher 1984 as *Glenodinium borgei*), SD (Hern et al. 1979c), VA (Parson and Parker 1989), WI (Prescott 1951b)

Peridiniopsis cunningtonii
GA (F Morris et al. 1978), OH (Carty 1993), VA (Parson and Parker 1989), WI (Prescott 1951b as *Glenodinium quadridens*)

Oaxaca (Figueroa-Torres and Moreno-Ruiz 2003)

Costa Rica (Haberyan et al. 1995)

Peridiniopsis edax
AZ (Taylor et al. 1979a), CA (M Morris et al. 1979b), CO (M Morris et al. 1979c), IL (M Morris et al. 1978a), KS (Thompson 1950), MT (Hern et al. 1979b), NC (M Morris et al. 1978b), NM (Lambou et al. 1979a), VA (Bovee 1960 as *Gymnodinium edax*)

Peridiniopsis elpatiewskyi
Greenland (Hansen 1967)

BC (Wailes 1934)

AZ (Taylor et al. 1979a), IN (Wailes 1934), MD (Thompson 1947), NJ (Williams et al. 1978), TN (Johansen et al. 2007)

MI (SC/VF L Antoine 8-19-03), NY (SC/VF Seneca L 8-18-97), OH (SC/VF Delta Reservoir 6-21-97)

Morelos (Figueroa-Torres and Moreno-Ruiz 2003)

Cuba (Popovský 1970), PR (Agreda 2006)

Peridiniopsis kulczynskii
AR (Hilgert et al. 1978a), KS (Thompson 1950), MN (Meyer and Brook 1969), MT (Morgan 1971), WI (Prescott 1951b)

Peridiniopsis oculatum

BC (Wailes 1934, Stein and Borden 1979)

AR (Hilgert et al. 1978a), AZ (Taylor et al. 1979a), CA (M Morris et al. 1979b), IA (Williams et al. 1979c), ID (F Morris et al. 1979c), IN (Wailes 1934), KS (Thompson 1950), LA (Lambou et al. 1979c), MS (Williams et al. 1977), MO (M Morris et al. 1979a), MT (Hilgert 1976), ND (Taylor et al. 1979b), NE (F Morris et al. 1979b), NM (Lambou et al. 1979a), OH (Wailes 1934), OK (Hern et al. 1979a), OR (Taylor et al. 1979c), SD (Hern et al. 1979c), TN (Forest 1954a), UT (Williams et al. 1979d), VA (Parson and Parker 1989 as *Glenodinium oculatum*)

Peridiniopsis penardii

ON (Duthie and Socha 1976)

CA (Horne et al. 1972, Herrgesell et al. 1976), IN (Wailes 1934), NC (M Morris et al. 1978b)

Peridiniopsis quadridens

ON (Johnson et al. 1968 as *Glenodinium quadridens*)

AR (Hilgert et al. 1978a), AZ (Taylor et al. 1979a), CA (M Morris et al. 1979b), IA (Prescott 1927), ID (F Morris et al. 1979c), IL (Eddy 1930), IN (Wailes 1934), KS (Williams et al. 1979b), KY (Kobraei and White 1996), LA (Lambou et al. 1979c), MO (M Morris et al. 1979a), MT (Morgan 1971), ND (Taylor et al. 1979b), NM (Lambou et al. 1979a), OH (Taft and Taft 1971, Carty 1993), OK (Vinyard 1958), TN (Johansen et al. 2007), TX (Carty 1986, 1989), VA (Forest 1954b), WI (Eddy 1930)

FL (SC 031208 L Michelle), NY (JH2008.09.21-2), VT (SC080731-7)

Cuba (Popovský 1970 as *Peridinium cunningtonii*), PR (JH2009.01.25-11)

Panama (Prescott 1951a)

Peridiniopsis thompsonii

IL (Eddy 1930 as *Peridinium cunningtonii*, Tiffany and Britton 1952), MD (Thompson 1947), TN (Forest 1954a)

NY (SC/VF Seneca L 8-18-97

Peridinium aciculiferum

MB (Wailes 1928a, Hecky et al. 1986), NS (Hughes 1947–1948 as *Glenodinium aciculiferum*), NT (Sheath and Steinman 1982), ON (Duthie and Socha 1976), PEI (Hughes 1947–1948 as *Glenodinium aciculiferum*), QC (Janus and Duthie 1979)

AK (Hilliard 1959), IA (Williams et al. 1979c), MO (M Morris et al. 1979a), NC (Whitford and Schumacher 1969), ND (Phillips and Fawley 2002), NH (Gruendling and Mathieson 1969), OH (Taft and Taft 1971 as *Glenodinium aciculiferum*), SC (Goldstein and Manzi 1976), VA (Woodson and Seaburg 1983)

Tabasco (Figueroa-Torres and Moreno-Ruiz 2003)

Panama (Prescott 1951a)

Peridinium bipes

BC (Wailes 1934), ON (Duthie and Socha 1976), QC (Bourrelly 1966)

AK (Prescott and Vinyard 1965), AR (Meyer 1969), CA (M Morris et al. 1979b), IL (Tiffany and Britton 1952), MA (Croasdale 1948), MN (Meyer and Brook 1969), NC (Whitford and Schumacher 1969), OH (Carty 1993), OK (Vinyard 1958, Pfiester and Terry 1978), VA (Marshall and Burchardt 2004)

Panama (Prescott 1955)

WV (JH2007.03.25)

Peridinium cinctum

Greenland (Hansen 1967)

BC, MB (Wailes 1934), NL (Davis 1973, Duthie et al. 1976), NS (Hughes 1947–48), ON (Duthie and Socha 1976), NT (Sheath and Steinman 1982), NU (Whelden 1947), QC (Wailes 1934), SK (Hammer et al. 1983)

AK (Alexander et al. 1980), AR (Meyer 1969), CO (Durrell and Norton 1960), FL (Van Meter-Kasanof 1973), IA (Prescott 1927), IL (Eddy 1930, Tiffany and Britton 1952), IN (Wailes 1934), KY (Cole 1957), MA (Webber 1961), MD (Thompson 1947), MI (Prescott 1951b), MN (Meyer and Brook 1969), MT (Morgan 1971), NC (Whitford and Schumacher 1969), ND (Phillips and Fawley 2002), NH (Gruendling and Mathieson 1969), NY (Joyce 1936), PA (Herbst and Hartman 1981), SC (Goldstein and Manzi 1976), SD (Hern et al. 1979c), TN (Eddy 1930, Forest 1954a), TX (Carty 1986), VA (Woodson and Seaburg 1983), WI (Eddy 1930, Prescott 1951b)

Jalisco (Figueroa-Torres and Moreno-Ruiz 2003), Michoacán (Tafall 1941), Morelos, Oaxaca, Tamaulipas (Figueroa-Torres and Moreno-Ruiz 2003)

Peridinium cinctum var tuberosum

NL (Duthie et al. 1976), QC (Puytorac et al. 1972)

MT (Morgan 1971), NC (Whitford and Schumacher 1969), NH (Yeo 1971), VA (Woodson and Seaburg 1983), WI (Prescott 1951b)

OH (SC Bowling Green)

Peridinium cinctum var westii

IA (Prescott 1927), NC (Whitford and Schumacher 1969)

Peridinium gatunense

ON (Duthie and Socha 1976), QC (Puyorac et al. 1972), SK (Hammer et al. 1983)

FL (Van Meter-Kasanof 1973), IL (Wunderlin 1971), MA (Prescott and Croasdale 1937 as *P. cinctum*), MN (Meyer and Brook 1969), MT (Morgan 1971), OH (Carty 1993), OK (Pfiester 1977), PA (Seaborn et al. 1992), TN (Johansen et al. 2007), TX (Carty 1986), VA (Forest 1954b), WI (Prescott 1951b)

AK (SC060713-5, Juneau), NC (SC090605-3 Charlotte), NY (JH2008.07.18-15), WV (JH2006.10.22-3)

Cuba (Popovský 1970), Panama (Ostenfeld and Nygaard 1925), PR (JH2009.01.25-2)

Mexico: San Luis Potosi (Cantoral-Uriza and Montejano-Zurita 1993), Yucatan (Krakhmalny et al. 2006)

Guadeloupe (French Antilles) (Couté and Tell 1990)

Peridinium godlewskii
KS (Thompson 1950), NC (Whitford and Schumacher 1984)

Peridinium gutwinskii
Panama (Prescott 1951a), PR (JH2009.01.25-2)

Peridinium keyense
NC (Whitford and Schumacher 1984)

Peridinium limbatum
BC (Stein 1975), NL (Duthie et al. 1976), NB, NS (Hughes 1947–1948), ON (Duthie and Socha 1976, Yan and Stokes 1978), QC (Bourrelly 1966)

GA (Schumacher 1956), MA (Evitt and Wall 1968), ME (Davis et al. 1978), MS (Canion and Ochs 2005), NH (Gruendling and Mathieson 1969), NC (Whitford and Schumacher 1969), NY (Martin 2006), OK (Pfiester and Skvarla 1980),VA (Marshall 1976), WI (Prescott 1951b)

MI (SC/VF Watersmeet 8-19-03), MS (AC 2004), NY (JH2009.05.11-9), PA (JH2008.05.24-1)

Peridinium lomnickii
BC (Dickman 1968)

CO (Toetz and Windell 1993), ID (F Morris et al. 1979c), TN (Johansen et al. 2007)

AR (SC030312-4)

Estado de Mexico (Figueroa-Torres and Moreno-Ruiz 2003)

Peridinium raciborskii
BC (Dickman 1968), NL (Duthie et al. 1976), ON (Schindler and Holmgren 1971, Duthie and Socha 1976), QC (Bourrelly 1966)

AR (Meyer 1969), MN (Meyer and Brook 1969)

Peridinium subsalsum
Cuba (Popovský 1970)

Peridinium volzii
BC (Wailes 1934), NL (L: Duthie et al. 1976), NS (Hughes 1947–48), NT (Sheath and Steinman 1982), ON (Kling and Holmgren 1972)

AK (Hilliard 1959), CO (Whitford and Kim 1971), FL (Van Meter-Kasanof 1973), MA (Prescott and Croasdale 1937), MD (Thompson 1947), MN (Eddy 1930,

Meyer and Brook 1969), MS (Canion and Ochs 2005), MT (Sieminska 1965), NC (Whitford and Schumacher 1969), OH (Carty 1993), OK (Vinyard 1958, Pfiester and Skvarla 1979), TN (Johansen et al. 2007), TX (Carty 1986), VA (Parson and Parker 1989)

CA (JH2008.08.15-5)

Costa Rica (Haberyan et al. 1995), Panama (Prescott 1951a)

Peridinium volzii f. vancouverense

AB (Thomasson 1962), NL (Duthie et al. 1976), ON (Bourrelly 1966)

Peridinim wierzejskii

FL (SC 031208 L Michelle), MI (SC08-19-03, Watersmeet), WI (Gina LaLibere)

Peridinium willei

Greenland (Hansen 1967)

AB (Thomasson 1962), BC (Wailes 1934, Stein and Borden 1979), NL (Duthie et al. 1976), ON (Duthie and Socha 1976), NWT (Sheath and Steinman 1982), SK (Hammer et al. 1983), QC (Puytorac et al. 1972)

AK (Wailes 1934), AR (Meyer 1969), AZ (Taylor et al. 1979a), CA (Thomasson 1962), CO (Durrell and Norton 1960, Whitford and Kim 1971), IA (Prescott 1927), ID (F Morris et al. 1979c), IL (Wunderlin 1971), MD (Thompson 1947), MN (Meyer and Brook 1969), MT (Thomasson 1962), NC (Whitford and Schumacher 1969), NH (Gruendling and Mathieson 1969), OH (Carty 1993), OK (Vinyard 1958, Pfiester 1976), OR (Thomasson 1962), PA (Herbst and Hartman 1981), SD (Hern et al. 1979c), TN (Eddy 1930, Forest 1954a), TX (Carty 1986), UT (Stewart and Blinn 1976), VA (Parson and Parker 1989), WA (F Morris et al. 1979a), WI (Prescott 1951b), WY (Thomasson 1962)

CT (SC080729-5), MA (SC080730-3), VT (SC080731-7)

Michoacán (Tafall 1941), Guanajuato (Lopez-Lopez and Serna-Hernandez 1999), Veracruz (Figueroa-Torres and Moreno-Ruiz 2003)

Peridinium wisconsinense

ON (Nicholls 1973), NS (Hughes 1947–48), NWT (Sheath and Steinman 1982), QC (Brunel and Poulin 1992), NB (SC080803-3)

AL (Taylor et al. 1977), AR (Meyer 1969), GA (F Morris et al. 1978), LA (Lambou et al. 1979c), MA (Prescott and Croasdale 1937), MI (Prescott 1951b), MN (Eddy 1930, Meyer and Brook 1969), MS (Canion and Ochs 2005), MT (Morgan 1971), NC (Whitford and Schumacher 1969), ND (Phillips and Fawley 2002), NH (Gruendling and Mathieson 1969), NJ (Williams et al. 1978), PA (Seaborn et al. 1992), RI (Hanisak 1973), SC (Goldstein and Manzi 1976), TN (Johansen et al. 2007), TX (Carty 1986), VA (Marshall 1976, Parson and Parker 1989), WI (Eddy 1930), WV (Perez et al. 1994)

CT (SC080729-1), ME (SC080803-1), NY (JH2008.07.18-15), VT (SC080731-6)

Piscinoodinium limneticum

Mexico (Ortega 1984)

Prorocentrum compressa
  MD (Thompson 1950)

Prosoaulax lacustre
  BC (Wailes 1928a, 1934)

Pseudoactiniscus apentasterias
  NU (Bursa 1969)

Rufusiella insignis
  KS (Richards 1962), MT (Morgan 1971 as *Urococcus insignis* but listed under Pyrrhophyta)
  TN (Johansen et al. 2007), VA (Forest 1954b as *Urococcus insignis* but listed under Pyrrhophyta)
  OH (Jeff Johansen, Cuyahoga Park, Cleveland)

Sphaerodinium polonicum
  IL (Eddy 1930, Tiffany and Britton 1952 as *Glenodinium cinctum*), KS (Thompson 1950), MN (Meyer and Brook 1969 as *Sphaerodinium cinctum*), VA (Parson and Parker 1989 as *Glenodinium cinctum*), WI (Eddy 1930 as *Glenodinium cinctum*)
  Distrito Federal (Ortega 1984), Yucatan (Schmitter-Soto et al. 2002)
  Belize (Carty and Wujek 2003)

Sphaerodinium fimbriatum
  KS (Thompson 1950), OK (Vinyard 1958), TX (Carty 1986)
  OH (SC970627 Pioneer)

Staszicella dinobryonis
  QC (Bourrelly 1966)
  UT (Rushforth and Squires 1985)

Stylodinium globosum
  BC (Stein and Borden 1979)
  MD (Thompson 1947), MN (Meyer and Brook 1969), OH (Taft and Taft 1971, Carty 1993), OK (Pfiester and Popovský 1979)
  TN (SC040518-2), TX (SC 8-24-82)
  Guadeloupe (Bourrelly and Manguin 1952)
  Panama (Prescott 1951a)

Stylodinium longipes
  MD (Thompson 1947)

Tetradinium javanicum
  BC (Stein 1975), Labrador (Duthie et al. 1976 as *Tetradinium intermedium*), ON (Duthie and Socha 1976 as *Tetradinium minus*)

MD (Thompson 1947), MI (Prescott 1951b), MN (Meyer and Brook 1969), OH (Taft and Taft 1971, Carty 1993), WI (Prescott 1951b)

TX (SC 8-24-82)

Guadeloupe (Bourrelly and Manguin 1952 as *Tetradinium intermedium*)

Tetradinium simplex
MI (Prescott et al. 1949)

Thompsodinium intermedium
KS (Thompson 1950), NC (Johansen et al. 2004), OH (Carty 1993), TN (Johansen et al. 2004), TX (Carty 1986, 1989)

Cuba (Popovský 1970), Belize (Carty and Wujek 2003)

FL (SC 950131 3A3), ME (SC 080801-3), MI (SC/VF Cedar Lake030820), NY (SC Laural Lake970814), WI (G LaLiberte 2005, Matthew Harp, Sunset Lake, 7-18-07)

Tovellia glabra
KS (Thompson 1950), OH (Klarer et al. 2000), TX (Carty 1986)

FL (SC070918 L Michelle), MT (SC030816-6), WI (SC030818-7)

Tyrannodinium berolinense
KS (Thompson 1950), WI (Wedemayer and Wilcox 1984)

Woloszynskia cestocoetes
MD (Thompson 1947), WI (Graham et al. 2004)

Woloszynskia ordinata
NWT (Sheath and Steinman 1982), ON (Kling and Holmgren 1972)

AR (Hilgert et al. 1978a), AZ (Taylor et al. 1979a), CA (M Morris et al. 1979b), IA (Williams et al. 1979c), IL (M Morris et al. 1978a), KS (Williams et al. 1979b), LA (Lambou et al. 1979c), MD (Thompson 1947), MT (Hern et al. 1979b), MS (Williams et al. 1977), NC (M Morris et al. 1978b), NE (F Morris et al. 1979b), NV (Lambou et al. 1979b), OH (Hilgert et al. 1978b), OK (Vinyard 1958), SD (Hern et al.1979c), TN (Forest 1954a), WY (Williams et al. 1979a)

Woloszynskia pascheri
NWT (Sheath and Steinman 1982 as *Gymnodinium veris*), ON (Gerrath and Nicholls 1974 as *Gymnodinium pascheri*)

Woloszynskia reticulata
KS (Thompson 1950), OK (Pfiester et al. 1980), TX (Carty 1986)

FL (SC070918, L. Michelle), OH (SC8-8-1997 Big Island)

Woloszynskia vera
NU (Kalff et al. 1975)

Appendix B
Geographic References by Location

References consulted but containing no dinoflagellate reports are cited in full. References that include reports on dinoflagellates and are in the literature cited are listed with an author-date citation.

CANADA

Yung, Y.K., P. Stokes, and E. Gorham. 1986. Algae of selected continental and maritime bogs in North America. Can. J. Bot. 64:1825–1833.

Alberta

Hickman, M. 1979. Seasonal succession, standing crop and determinants of primary productivity of the phytoplankton of Ministik Lake, Alberta, Canada. Hydrobiologia 64:105–121.
Thomasson 1962.

British Columbia

Lowe 1931.
Stein and Borden 1979.
Wailes 1928a, 1934.

Manitoba

Hecky et al. 1986.
Shamess et al. 1985.

New Brunswick

Hughes 1947–1948.

Newfoundland and Labrador

Davis, C.C. 1972. Plankton dynamics in a Newfoundland lake. Verh. Int. Verein. Theor. Angew. Limnol. 18:278–283.
Davis 1972, 1973.

Duthie et al. 1976.
Palmer, C.M. 1965. Phytoplankton periodicity in a Newfoundland pond. J. Phycol. 1:38–39.

Northwest Territories

Bursa 1969.
Findlay 2003.
Moore, J.W. 1978. Distribution and abundance of phytoplankton in 153 lakes, rivers, and pools in the Northwest Territories. Can. J. Bot. 56:1765–1773.
Sheath and Hellebust 1978.
Sheath, R.G., and J.A. Hellebust. 1982. Effects of long-term natural acidification on the algal communities of tundra ponds at the Smoking Hills, N.W.T., Canada. Can. J. Bot. 60 (1): 58–72.
Sheath and Steinman 1964.
Sheath, R.G., M. Munawar, and J.A. Hellebust. 1975. Fluctuations of phytoplankton biomass and its compostion in a subarctic lake during summer. Can. J. Bot. 53:2240–2246.

Nova Scotia

Smith 1961.

Nunavut

Bursa 1969.
Kalff et al. 1975.
Whelden 1947.

Ontario

Duthie and Socha 1976.
Findlay 2003.
Johnson et al. 1968.
Kling and Holmgren 1972.
Nicholls 1998.

Prince Edward Island

Hughes 1947–1948.

Quebec

Bourrelly 1966.
Dickman, M., and M. Johnson. 1975. Phytoplankton of five lakes in Gatineau Park, Quebec. Can. Field-Nat. 89:361–370.
Lowe 1927.
Plinski, M., and E. Magnin. 1979. Analyse écologique du phytoplancton de tros lacs des Laurentides (Québec, Canada). Can. J. Bot. 57:2791–2799.

Saskatchewan

Hammer et al. 1983.

Kuehne, P. 1941. The phytoplankton of southern and central Saskatchewan. Part 2. Can. J. Res. C 19(9):313–322.

Lund, J.W.G. 1962. Phytoplankton from some lakes in northern Saskatchewan and from Great Slave Lake. Can. J. Bot. 40:1499–1514.

Yukon Territory

Hooper 1947.

CENTRAL AMERICA AND THE CARIBBEAN

Bahamas

Britton, N.L., and C.F. Millspaugh. 1920. The Bahama flora. Reprint, New York Botanical Garden, 1962. [algae but no dinoflagellates]

Barbados, Dominica

Almodovar, L.R., and F.A. Pagan. 1967. Notes on the algae of Barbados. Nova Hedwigia 13:11–115. [marine]

West, G.S. 1904. West Indian freshwater algae. J. Bot. 42:281–294.

Belize

Carty and Wujek 2003.

Bermuda

Britton, N. 1918. Flora of Bermuda. Reprint, Hafner, NY, 1965. [algae but all marine]

Collins, F.S., and A.B. Hervey. 1917. The algae of Bermuda. Proc. Am. Acad. Arts Sci. 53 (1). [freshwater algae but no dinoflagellates]

Howe, M.A. 1924. Notes on algae of Bermuda and the Bahamas. Bull. Torrey Bot. Club 51:351–359, also Contrib. N.Y. Bot. Gard. 258. [no dinoflagellates]

Costa Rica

Haberyan et al. 1995.

Hargraves and Víquez 1981.

Umana, G.V. 2001. Limnology of Botos Lake, a tropical crater lake in Costa Rica. Rev. Biol. Trop. 49 (Suppl. 2): 1–10.

Umaña Villalobos 1988.

Cuba

Laíz, O., et al. 1993a. Limnology of Cuban reservoirs I. Lebrije. Trop. Freshw. Biol. 3:371–396.

Laíz, O., et al. 1993b. Limnology of Cuban reservoirs: II. Higuanajo. Acta Cient. Venez. 44:297–306.

Laíz, O., et al. 1994. Comparative limnology of four Cuban reservoirs. Int. Rev. Gesamt. Hydrobiol. 79:27–45.

Margalef 1947.

Popovský 1970.

French Antilles (Guadeloupe, Martinique, St. Martin, St. Barts)

Bourrelly and Manguin 1952.

Couté and Tell 1990.

Mazé, H., and A. Schramm. 1870–1877. Essai de classification des algues de la Guadaloupe. Imprimérie du Gouvernement, Basse Terre.

Starmuhlner, F., and Y. Therezien. 1983. Résultats de la Mission Hydrobiologique Austro-Française de 1979 aux Iles de la Guadeloupe, de las Dominique et de las Martinique (Petites Antilles). Ann. Naturhist. Mus. Wien. 85b: 219–262.

Guatemala

Clark, H.W. 1908. The holophytic plankton of Lakes Atitlan and Amatitlan, Guatemala. Proc. Biol. Soc. Wash. 21:91–106. (Ch, *Peridinium tabulatum*)

Ostenfeld and Nygaard 1925.

Taylor, W.R. 1939. Fresh-water algae from the Petén district of Guatemala. Bot. Not. 1939:112–124. [no dinoflagellates]

Tilden, J.E. 1908. Notes on a collection of algae from Guatemala. Proc. Biol. Soc. Wash. 21:153–156.

Jamaica

Hegewald, E. 1976. A contribution to the algal flora of Jamaica. Nova Hedwigia 28:45–69.

Mexico

Cantoral-Uriza, E.A., and G. Montejana-Zurita. 1993. Algae from El Sulto, San Luis Potosi, Mexico: an example of floristic studies in changing environment. Bol. Soc. Bot. Mex. 53:3–20.

Figueroa-Torres, G., and J.L. Moreno-Ruiz. 2003. Dinoflagelados dulceacuícolas de México. In T. Barreiro-Güemes, E.M. del Castillo, M. Signoret-Poillon, and G. Figueroa-Torres, eds., Planctología Mexicana. Sociedad Mexicana de Planctología, Baja California Sur, México, 85–102.

Lopez-Lopez and Serna-Hernandez 1999.

Ortega, M.M. 1984. Catálogo de algas continentales recientes de México. Univ. Nac. Autón. de México.

Schmitter-Soto, J.J., et al. 2002. Hydrogeochemical and biological characteristics of cenotes in the Yucatan Peninsula (SE Mexico). Hydrobiologia 467:215–228.

Panama

Ostenfeld and Nygaard 1925.

Prescott, G.W. 1936. Notes on the algae of the Gatun lake, Panama Canal. Trans. Am. Microsc. Soc. 55:501–509.

Prescott 1951a.

Puerto Rico

Agreda 2006.

Britton, M.E. 1937. The freshwater algae of Puerto Rico. MS thesis, Ohio State University. [no dinoflagellates]

Foerster 1971.

Hernández-Becerril, D.U., and M. Navarro. 1996. Thecate dinoflagellates (Dinophyceae) from Bahia Fosforescente, Puerto Rico. Rev. Biol. Trop. 44:465–475. [marine]

Lagerheim, G. 1887. Ueber einige Algen aus Cuba, Jamaica, und Puerto Rico. Bot. Not. 1887:193–199.

Tiffany, L.H. 1936. Wille's collection of Puerto Rican freshwater algae. Brittonia 2:165–175.

Wille, N. 1915. Report on an expedition to Porto Rico for collecting fresh-water algae. J. N.Y. Bot. Gard. 16:132–146.

GREENLAND

Børgesen, F. 1910. III. Freshwater algae from the Danmark-expedition to north-east Greenland (N. of 76°N. lat.). Medd. Grønland 43:69–90.

Hansen 1967.

Larsen, E. 1904. II. The freshwater algae of East Greenland. Medd. Grønland 30:77–110.

Moestrup et al. 2008.

Moss, E.L. 1879. Preliminary notice of the surface-fauna of the Arctic Seas, as observed in the recent Arctic Expedition. J. Linn. Soc. Zool. 14:122–126. [marine dinoflagellates]

Peterson, J.B. 1924 (1927?). XIII. Freshwater algae from the north coast of Greenland collected by the late Dr. TH. Wulff. Medd. Grønland 64:305–319.

Vanhöffen, E. 1897. Peridineen und Dinophyceen. In: Botanische Ergebnisse. Dr. v. Drygalski's ausgesandten Grönlandexpedition. Biblio. Bot. 42:25–27. [marine dinoflagellates and freshwater *Dinobryon*]

UNITED STATES

Alabama

Lackey 1936, Taylor et al. 1977.

Alaska

Alexander et al. 1980.

Croasdale, H.T. 1958. Freshwater algae of Alaska 2. Some new forms from the plankton of Karluk Lake. Trans. Am. Microsc. Soc. 77:31–35.

Hilliard 1959.

Hooper 1947.

Kol, E. 1942. The snow and ice algae of Alaska. Smithson. Misc. Coll. 101 (16): 1–36, 6 plates.

Prescott, G.W. 1963. Ecology of Alaskan freshwater algae II. Introduction: general considerations. Trans. Am. Microsc. Soc. 82:83–98. [mentioned but no genera given]

Prescott and Vinyard 1965.

Wailes 1934.

Arizona

Cameron 1964, Taylor et al. 1979a.

Arkansas

Hilgert et al. 1978a.
Kelley and Pfiester 1989.
Meyer 1969.

California

Allen 1920.
Herrgesell et al. 1976.
Horne et al. 1972.
M.K. Morris et al. 1979b.
Thomasson 1962.

Colorado

Bursa 1970.
Durrell and Norton 1960.
Kolesar et al. 2002.
M.K. Morris et al. 1979c.
Toetz and Windell 1993.

Connecticut

Conn 1905.
Huszar et al. 2003.

Florida

Dawes, C.J., and J. Jewett-Smith. 1985. Algal flora of an acid marsh in west-central Florida. Trans. Am. Microsc. Soc. 104:188–193.
Taylor et al. 1978.
Van Meter-Kasanof 1973.
Weldon, R. 1941. Some observations of freshwater algae of Florida. J. Elisha Mitchell Sci. Soc. 57:262–272.

Georgia

F.A. Morris et al. 1978.
Schumacher 1956.

Idaho

F.A. Morris et al. 1979c.

Illinois

Kofoid 1908.
M.K. Morris et al. 1978a.
Tiffany and Britton 1952.
Wunderlin 1971.

Indiana

Cook, G.W. 1951. The phytoplankton of two artificial lakes in Hendricks County, Indiana. Butler Univ. Bot. Stud. 10:53–60.
Daily and Miner 1953.

Iowa

Prescott 1927.
Williams et al. 1979c.

Kansas

McFarland, H.J., E.A. Brazda, and B.H. McFarland. 1964. A preliminary survey of the algae of Cheyenne Bottoms in Kansas. Fort Hays Stud. 2:1–80.
Richards 1962.
Thompson 1938, 1949, 1950.
Williams et al. 1979b.

Kentucky

Cole 1957.
Dillard, G.E., and S.B. Crider. 1970. Kentucky algae. I. Trans. Ky. Acad. Sci. 31: 66–72.
Dillard et al. 1976.
Geiling, W., and L. Krumholz. 1964. A limnological survey of sink-hole ponds in the vicinity of Doe Run, Meade County, Kentucky. Trans. Ky. Acad. Sci. 24:37–80.
Kobraei and White 1996.
McInteer, B.B. 1932. A survey of the algae of Kentucky. PhD dissertation, Ohio State University.

Louisiana

Farmer and Roberts 1990.
Lambou et al. 1979c.
Prescott, G.W. 1942. The algae of Louisiana, with descriptions of some new forms and notes on distribution. Trans. Am. Microsc. Soc. 61:109–119.

Maine

Davis et al. 1978.

Maryland

Bold, H.C. 1938. Notes on Maryland algae. Bull. Torrey Bot. Club 65:293–301.
Thompson 1947, 1949, 1950, Thompson and Meyer 1984.

Massachusetts

Auyang 1962.
Croasdale, H.T. 1935. The fresh water algae of Woods Hole, Massachusetts. PhD dissertation, University of Pennsylvania, Philadelphia.
Herdman 1924.
Prescott and Croasdale 1937.
Prescott, G.W., and H.T. Croasdale. 1942. The algae of New England—II. Additions to the freshwater algal flora of Massachusetts. Am. Midl. Nat. 27:662–676.
Thompson 1949.
Webber 1961.
Wright 1964.

Michigan

Carrick 2005.
Prescott 1951b.
Sicko-Goad and Walker 1979.
Welch, P.S. 1936. A limnological study of a small *Sphagnum*-leatherleaf–black spruce bog lake with special reference to its plankton. Trans. Am. Microsc. Soc. 55:300–312.

Minnesota

Meyer and Brook 1969.

Mississippi

Canion and Ochs 2005.
Whitford, L.A. 1950. Some freshwater algae from Mississippi. Castanea 15:117–123.
Williams et al. 1977.

Missouri

Fields and Rhodes 1991.
Hayden, A. 1910. The algal flora of the Missouri Botanical Garden. Annu. Rep. Mo. Bot. Gard. 1:25–48.
M.K. Morris et al. 1979a.

Montana

Garric, R. 1962. A study of the cryoflora in the Pacific Northwest states. MS thesis, Montana State University.
Hern et al. 1979b.
Hilgert 1976.
Hoham 1966.
Kidd, D.E. 1964. A quantitative analysis of phytoplankton along a Rocky Mountain Divide transect. Trans. Am. Microsc. Soc. 83:409–420.
Morgan 1971.
Sieminska 1965.
Thomasson 1962.
Vinyard 1957.

Nebraska

F.A. Morris et al. 1979b.

Nevada

Lambou et al. 1979b.
LaRivers, I. 1978. Algae of the western Great Basin. Bioresources Center, Desert Research Institute, University of Nevada. [390 pages and not one dinoflagellate!]

New Hampshire

Gruendling and Mathieson 1969.
Yeo 1971.

New Jersey

Barlow and Triemer 1988.
Williams et al. 1978.

New Mexico

Lambou et al. 1979a.

New York

CSLAP 2003.
Huszar et al. 2003.
Joyce 1936.
Martin 2006.
Schumacher 1961.

North Carolina

Bellis, V. 1988. Microalgae of barrier island freshwater ponds, Nags Head, North Carolina. Assoc. Southeast. Biol. Bull. 35:164–169.
M.K. Morris et al. 1978b.
Whitford and Schumacher 1969, 1984.

North Dakota

Anderson, E.M., and E.R. Walker. 1920. An ecological study of the algae of some sandhill lakes. Trans. Am. Microsc. Soc. 39:51–85.
Moore, G.T. 1917. Preliminary list of algae in Devils Lake, North Dakota. Ann. Mo. Bot. Gard. 4:293–303.
Moore, G.T., and N. Carter. 1923. Algae from lakes in the northeastern part of North Dakota. Ann. Mo. Bot. Gard. 10:393–422.
Phillips and Fawley 2002.
Taft, C.E. 1948. Some algae from the Black Hills of South Dakota and the Turtle Mountain Region of North Dakota. Ohio J. Sci. 48 (2): 84–88.
Taylor et al. 1979b.

Ohio

Carty 1993.
Chapman, F.B. 1934. The algae of the Urbana (Ohio) raised bog. Ohio J. Sci. 34: 327–332.
Hilgert et al. 1978b.
Klarer et al. 2000.
Taft and Taft 1971.

Oklahoma

Hern et al. 1979a.
Nolen et al. 1989.
Pfiester and Lynch 1980.
Pfiester and Popovský 1979.
Pfiester et al. 1984.
Vinyard 1958.

Oregon

Taylor et al. 1979c.
Thomasson 1962.

Pennsylvania

Murphy et al. 1994.
Seaborn et al. 1992.

Rhode Island

Hanisak 1973.

South Carolina

Dillard 1967.
Goldstein and Manzi 1976.
Jacobs 1968, 1971.

South Dakota

Hern et al. 1979c.
Hudson, P.L., and B.C. Cowell. 1966. Distribution and abundance of phytoplankton and rotifers in a main stem Missouri River reservoir. Proc. South Dakota Acad. Sci. 45:84–106.
Taft, C.E. 1948. Some algae from the Black Hills of South Dakota and the Turtle Mountain Region of North Dakota. Ohio J. Sci. 48 (2): 84–88.

Tennessee

Eddy, S. 1930. The plankton of Reelfoot Lake, Tennessee. Trans. Am. Microsc. Soc. 49:246–251.
Forest 1954a.
Hill 1980.
Johansen et al. 2007.
Silva, H., and A.J. Sharp. 1944. Some algae of the southern Appalachians. J. Tenn. Acad. Sci. 19: 337–345.

Texas

Carty 1986.

Utah

Felix, E.A., and S.R. Rushforth. 1979. The algal flora of the Great Salt Lake, Utah, U.S.A. Nova Hedwigia 31:163–193+.
Williams et al. 1979d.

Vermont

Williams et al. 1979d.

Virginia

Bovee 1960.

Forest 1954b.

Marshall 1976.

Marshall and Burchardt 2004.

Parson and Parker 1989.

Woodson, B.R., and V. Holoman. 1964. A systematic and ecological study of algae in Chesterfield County, Virginia. Va. J. Sci. 15:52–70. [no dinoflagellates]

Woodson and Seaburg 1983.

Washington

Garric, R. 1962. A study of the cryoflora in the Pacific Northwest states. MS thesis, Montana State University.

Larson et al. 1998.

F.A. Morris et al. 1979a.

Schumacher and Muenscher 1952.

Smith, M.A., and M.J. White. 1985. Observations on lakes near Mount St. Helens: phytoplankton. Arch. Hydrobiol. 104:345–362.

West Virginia

Bennett, H.D. 1969. Algae in relation to mine water. Castanea 34:306–329.

Fling, E.M. 1939. One hundred algae of West Virginia. Castanea 4:11–25.

Perez et al. 1994.

Woodson 1969.

Wisconsin

Eddy 1930.

Prescott 1951b.

Wedemayer and Wilcox 1984.

Wedemayer et al. 1982.

Wilcox and Wedemayer 1984.

Woelkerling, W.J. 1976. Wisconsin desmids I. Aufwuchs and plankton communities of selected acid bogs, alkaline bogs, and closed bogs. Hydrobiologia 48:209–232. ["Dinophyta" but no genera]

Wyoming

Thomasson 1962.

Williams et al. 1979a.

Appendix C
Latin Diagnoses and Other Technical Issues

Actiniscus canadensis Bursa ex Carty

SYNONYM: *Actiniscus canadensis* Bursa 1969. J. Protozool 16:414.

LECTOTYPE: Fig. 17, pentasters Figs. 15 and 16 in Bursa 1969. J. Protozool 16:415.

Dinoflagellata aqua dulcis, achromatica, cellula sine paries, pentaster parvus siliceous prope capsulam nucleum, diameter 28–43 µm.

Freshwater, colorless, athecate cell with small siliceous pentasters near the nuclear capsule, diameter 28–43 µm.

COMMENTS: Latin diagnosis missing from original description. Original description cites Figs. 15–18, 20–22. Figures 15 and 16 are isolated, individual pentasters. Figure 17 is a ventral view of *A. canadensis;* Figure 18 is a disrupted cell of *A. canadensis.* Figure 20 is two pentasters. Figure 21 (no legend) is a ventral view of a cell, identical (though darker) than Fig. 19, which is identified as *Pseudoactiniscus apentasterias.* Figure 22 (no legend) is of cell(s) undergoing division or fusion.

Peridiniopsis thompsonii Bourrelly ex Carty

SYNONYM: *Peridiniopsis thompsonii* Bourrelly 1968a. Protistologica 4:9.

LECTOTYPE: Pl. 2, Figs. 12–15 in Thompson 1947, as *Glenodinium quadridens.* State of Maryland Board of Natural Resources 67:14.

Dinoflagellata aqua dulcis, chloroplasto, formula kofoidiana thecarum porus apicalis, 5' 0a, 7", 5''', 2''''; parietorum postcingularum et antapicalis spinis.

COMMENTS: Latin diagnosis missing. Bourrelly recognized that the cell illustrated by Thompson was not *Peridiniopsis quadridens* and named it for Thompson.

Pseudoactiniscus Bursa ex Carty

BASIONYM: *Pseudoactiniscus* Bursa 1969. J. Protozool 16:414.

TYPE SPECIES: *Pseudoactiniscus apentasterias* Bursa 1969. J. Protozool 16:414.

Dinoflagellata aqua dulcis, sphaerica, sine chloroplast, sine muris, sine pentaster, nucleus ad centralus, cysta rotunda.

Freshwater, round, colorless, athecate dinoflagellate with central nucleus; no pentasters; spherical cysts.

Pseudoactiniscus apentasterias Bursa ex Carty

SYNONYM: *Pseudoactiniscus apentasterias* Bursa 1969. J. Protozool 16:414.

LECTOTYPE: Fig. 19 from Bursa 1969. J. Protozool 16:414–415.

Cellula aqua dulcis, sphaerica, sine chloroplast, sine muris, sine pentaster, nucleus ad centralus, cum guttae olei, Arctae.

Cells freshwater, round, colorless, athecate dinoflagellate with central nucleus; no pentasters; with oil bodies, found in the Arctic.

Dinamoeba coloradense (Bursa) Carty comb. nov.

BASIONYM: *Dinamoebidium coloradense* Bursa 1970. Arctic and Alpine Research 2 (2): 146, Figs. 1–8.

COMMENT: *Dinamoebidium* is considered a junior synonym of *Dinamoeba* (Fensome et al. 1993, 165).

Durinskia dybowski (Wołoszyńska) Carty comb. nov.

SYNONYM: *Peridinium dybowski* Wołoszyńska 1916. Polskie Peridineae słodkowodne—Polnische Süßwasser-Peridineen. Bull Acad. Sc. Cracovie Cl. Math. Nat. Sér. B. Sc. Nat., 273, Taf. 13, Figs. 9–14.

COMMENTS: Wołoszyńska (1916) described *Peridinium dybowski* as having six precingular plates and drew rows of punctate ornamentation on the plates. This species is sometimes considered a synonym of *Durinskia baltica*. *Durinskia baltica* was originally described as *Glenodinium balticum* Levander 1894, and the plate pattern came to be one of those proposed for the unknown tabulation of *Glenodinium cinctum* (Levander 1894, Loeblich III 1980) (also see *Sphaerodinium*). *Glenodinium balticum* was transferred to *Peridinium* by Lemmermann (1910), then to *Durinskia* by Carty and Cox (1986), since the six precingular plates do not match the seven of *Peridinium*. Levander described the species *balticum* as found in brackish water. Extensive research was done on a binucleate form (Tomas et al. 1973, Chesnick and Cox 1987, 1989) that came from brackish Salton Sea (CA) isolates. My collections in Texas, Tennessee, and Ohio were from the plankton of strictly freshwater environments, and I think the preferred name for the freshwater species is *D. dybowski,* reserving *D. baltica* for brackish or marine specimens. *D. dybowski* has been reported (as *Glenodinium dybowski*) from Canada and the United States (Eddy 1930, Whelden 1947, Sheath and Steinman 1982).

Ceratium hirundinella f. *tridenta* n. forma

HOLOTYPE: Figs. 52–56 herein.
ETYMOLOGY: Name based on the trident appearance of cell.

Body wide and short, long straight apical horn, antapical horn straight, may be at an obtuse angle with apical horn. Postcingular horns similar in length and downcurved. No other form has curved postcingular horns. Total length (apical horn to antapical horn) 283–309 µm, maximum width (between postcingular horns) 100–114 µm, body width 43–56 µm, body length 20–38 µm.

Naiadinium Carty gen. nov.

TYPE SPECIES: *Naiadinium polonicum* (Wołoszyńska) Carty comb. nov.
HOLOTYPE: *Peridinium polonicum* Wołoszyńska 1916. Bull Acad. Sci. Cracovie B, 271, Taf. 12, Figs. 1–10.
ETYMOLOGY: Naiads are water spirits in mythology, and this species is common in the water; *-dinium* is a common dinoflagellate suffix.

Dinoflagellata aqua dulcis, chloroplasto, formula kofoidiana thecarum porus apicalis, 4', 1-2a, 7", 5C, 5'", 2""; 1' spina crassa ad infimum.

Freshwater, photosynthetic dinoflagellate with an apical pore, plate tabulation pattern 4', 1-2a, 7", 5C, 5'", 2""; large 1' plate; strong dorsoventral compression.

Naiadinium biscutelliforme (Thompson) Carty comb. nov.

LECTOTYPE: Thompson 1950. Lloydia 13:293–4, Figs. 35–40.
BASIONYM: *Glenodinium gymnodinium* Penard var. *biscutelliforme* Thompson 1950. Lloydia 13:293–294, Figs. 35–40.

Naiadinium polonicum (Wołoszyńska) Carty comb. nov.

BASIONYM: *Peridinium polonicum* Wołoszyńska 1916. Bull Acad. Sci. Cracovie B, 271, Taf. 12, Figs. 1–10.
SYNONYMS: *Glenodinium gymnodinium* Penard 1891. Bull trav. Soc. Bot. Genève 6: 54; *Peridiniopsis gymnodinium* (Penard) Bourrelly 1968a; *Peridiniopsis polonicum* (Wołoszyńska) Bourrelly 1968a.

COMMENTS: This taxon, whose plate tabulation pattern was first illustrated by Wołoszyńska (1916) as *Peridinium polonicum,* with one apical intercalary plate, was considered possibly a synonym of *Glenodinium gymnodinium* Penard (which has no known plate pattern) by Bourrelly (1968a), who moved species into the genus *Peridiniopsis* that did not have the *Peridinium* pattern of 3' 2-3a, 7". He made two new combinations for this taxon, *Peridiniopsis gymnodinium* (Penard) Bourrelly and *Peridiniopsis polonicum* (Wołoszyńska) Bourrelly (1968a). Penard's description and illustration do not conform with the species identified as *P. polonicum* (Carty 1989). Thompson found a species that looked just like the *Peridinium polonicum* illustrated by Wołoszyńska, but it had two apical intercalary plates and Thompson called it *Glenodinium gymnodinium* var. *biscutelliforme*. This genus is distinctive but is neither *Peridinium* nor *Peridiniopsis*.

The genus *Peridinium* no longer contains any species with two apical intercalary plates; the Umbonatum group has been moved to *Parvodinium* (Carty 2008) and the Palatinum group to *Palatinus* (Moestrup et al. 2009). When Bourrelly established his categories for *Peridiniopsis* species there was none for 4' 1-2a, 7". Molecular analysis of nuclear-encoded LSU rDNA placed this taxon (as *Peridiniopsis polonicum*) remote from the type species for *Peridiniopsis*, *P. borgei* (Craveiro et al. 2009), providing further support for its separate standing.

N. polonicum, with a single apical intercalary plate as originally described by Wołoszyńska (1916), has been identified, and illustrated by Popovský (1970 from Cuba), by Suxena (1983 from India), and by EPA workers (1977–1979) who distinguished it from var. *biscutelliforme*.

N. biscutelliforme as illustrated by Thompson (1950) is much more commonly seen, and if illustrations of "*Peridinium polonicum*" accompany a report (Tafall 1941, Roset et al. 2002), two apical intercalary plates are seen. The importance of separating the two species is that *N. biscutelliforme* has special interest as a known toxin-producing freshwater dinoflagellate. *Naiadinium biscutelliforme* as *Peridinium polonicum* is responsible for fish kills in Japan (Hashimoto et al. 1968, Oshima et al. 1989) and Spain (Roset et al. 2002), but there is no documentation of toxicity in *N. polonicum*.

Glossary

accumulation body Yellow, orange, or red, round to oval structure, usually in the hypotheca

amphidinioid Pertaining to a cell divided by cingulum into small epicone/theca and larger hypocone/theca

amphiesma Vesicle beneath cell membrane that may contain plates

apiculate Slightly drawn out (at apex of cell)

aplanospore Nonmotile spore

aplanozygote Like *planozygote* but applied to immobile dinoflagellates

assimilative Stage of life during which cell takes in nutrients

athecate Lacking cellulose armor plates

autospore Nonmotile spore found in parental wall

basionym Base name of taxon, original description

binary fission Cell division in which one cell pinches into two, each with some original cell covering

bloom Large number of a single species; plankton tow usually yields only one species

carina Ridge of plate material

cingulum Groove around a dinoflagellate cell housing the transverse tinsel flagellum

complexum event In thecate dinoflagellates, division of a plate into two

compression Flattening in the dorsoventral plane

crescent Narrow lunate, with one side convex, the other concave

cryophile Cold loving

depression Flattening in the anterior posterior plane

desmoschisis Cell division in which each daughter receives half the parental theca and regenerates the missing half, for example *Ceratium*

ecdyse To leave the theca; cell sheds theca

eleutheroschisis Cell division in which the theca is shed before cell divides

elliptical Elongate oval, rounded poles

epicone Upper portion of cell in taxa without a cell wall

epineuston Water-air interface area

epitheca Portion of the dinoflagellate cell above the cingulum in thecate taxa

euplankton Open-water, free-floating plankton

eutrophic Enriched with organic nutrients

fimbriate Fringed, slightly jagged

flange An extension of thecal plate into ribbon

fusiform Smoothly elongate, tapered to points

holophytic Photosynthetic; plantlike nutrition

holozoic Ingests large particles; animal-like nutrition

horn Extension of cell body composed of thecal plates

hypnospore Resting spore

hypnozygote Result of fusion of gametes; thick-walled, nonmotile form; occurs after theca is shed, triangular cysts of *Ceratium*

hypocone Lower portion of cell in nonthecate taxa

hypotheca Portion of a cell below the cingulum in thecate taxa

karyogamy Fusion of haploid nuclei during sexual reproduction

kleptoplastids Plastids retained by phagotrophic dinoflagellates from photosynthetic prey species

laminated Layered, as in the gelatinous sheaths of *Gloeodinium* and *Rufusiella*

lentic Still water, as in ponds, lakes, reservoirs

lithic Mineral, inorganic material

littoral Near shore

list Winglike extension of plate margin

lotic Flowing water, as in streams, rivers

lunate Fat crescent; one side convex, the other concave, tapering to points

mastigonemes Hairlike structures connecting ensheathed transverse flagellum to striated strand

mixotrophic Nutrition consisting of both photosynthesis and heterotrophy

monophyletic Pertaining to a single ancestor and all its descendants

myzocytosis Production of a *peduncle* (feeding tube) from the sulcal area that sucks prey contents

oligotrophic Low-nutrient conditions

osmotrophy Uptake of dissolved material

ovate Egg-shaped, rounded bottom, smoothly rounded tapered upper section

ovoid Slightly elongated sphere, symmetrical in three dimensions

parietal Found at periphery of cell

peduncle Extrusive body part used to ingest

pentaster Starlike internal siliceous structure whose arms surround nuclear capsule

phagotrophy Ingestion of other organisms

periphyton Epiphytic on attached macrophytes

planozygote Immediate result of fusion of gametes; zygote within original thecal wall with wide suture bands, is motile with two flagella, large

plasmogamy Fusion of the cell contents of two cells, but not the nuclei

punctate Dot type of ornamentation

pusule Osmoregulatory organelle, usually with a system of canals, opening into sulcus

pyrenoid Proteinaceous body

pyriform Pear-shaped; rounded bottom, tapered top half

reticulate Type of ornamentation with small ridges on plate forming polygons, often enclosing a trichocyst pore

simplex event In thecate dinoflagellates, two plates fusing into one

spindle-shaped Angular elongate, tapering to points

spine Usually short, acute extension of thecal plate material, usually at border of plate

stigma Eyespot

stipe Stalk

striae Plate ornamentation consisting of lines of punctae, or ridges

sulcus Groove down the front of the cell; contains pore through which flagella emerge

suture Place where two plates touch

taxon (pl. taxa) Unspecified taxonomic category

thecate Cell with cellulose plates

thylakoids Photosynthetic membrane-bound compartments

tinsel flagellum Flagellar type with (long) hairlike mastigonemes; found in the cingular groove in dinoflagellates

trichocyst Ejectile organelle

tumid Slightly swollen

tycoplankton Microscopic algae found floating near shore, mixed in shoreline vegetation

vermiform Type of ornamentation, wormlike

whiplash flagellum Trailing flagellum without mastigonemes; propels cell, directed by sulcal groove

zoospore Motile cell

Literature Cited

Abé, T.H. 1981. Studies on the family Peridinidae. An unfinished monograph of the armoured dinoflagellata. Publ. Seto Mar. Biol. Lab. Spec. Publ. Ser. 6.

Agreda, F.P. 2006. Dinámica fisicoquímica y fitoplanctónica del embalse Guajataca, Puerto Rico. MS thesis, Univ. P.R.

Alexander, V., D.W. Stanley,, R.J. Daley, and C.P. McRoy. 1980. Primary producers. In J.E. Hobbie, Limnology of tundra ponds, Barrow, Alaska. US/IBP Synthesis Ser. 13. Dowden, Hutchinson & Ross, Stroudsburg, PA.

Allen, W.E. 1920. A quantitative and statistical study of the plankton of the San Joaquin River and its tributaries in and near Stockton, California, in 1913. Univ. Calif. Publ. Zool. 22:1–292.

Allman, G.J. 1855. Observations on *Aphanizomenon flos-aquae* and a species of Peridinea. Q. J. Microsc. Sci. 3:21–25.

Amberg, O. 1903. Biologische Notiz über den Lago di Muzzano. Forschungsberichte der Biologischen Station zu Plön X. Teil, 74–85.

Anissimova, N.V. 1926. Novye Peridineae, najdennye v solennych vodoemach Staroj Russy (Novgorod. Gub.) Russk. Gidrobiol. Žhur. 5:188–193.

Apstein, C. 1896. Das Süßwasserplankton. Methode und Resultate der quantitativen Untersuchung. Kiel und Leipzig. Verlag von Lipsius & Tischer.

Auyang, T.S. 1962. A survey of the algae of Lake Quinsigamond. Rhodora 64: 49–59.

Bachmann, H. 1911. Das Phytoplankton des Süsswassers mit besonderer Berücksichtigung des Vierwaldstättersees. Literaturverzeichnis Jena, G. Fischer.

Bachmann, H. 1921. Beiträge zur Algenflora des Süsswassers von Westgrönland. Mitt. Naturforsch. Ges. Luzern. 8:1–181.

Bailey, J.W. 1850. Microscopical observations made in South Carolina, Georgia and Florida. Smithson. Contr. Knowl. 2:1–48.

Balech, E. 1951. Deuxième contribution à las connaissance des *Peridinium*. Hydrobiologia 3(4): 305–330.

Balech, E. 1974. El genero *Protoperidinium* Bergh, 1881 (*Peridinium* Ehrenberg, 1831, partim). Rev. Mus. Argent. Cienc. Nat. Bernardino Rivadavia 4 (1). Buenos Aires.

Balech, E. 1980. On thecal morphology of dinoflagellates with special emphasis on circular [*sic*] and sulcal plates. An. Cent. Cienc. Mar Limnol. Univ. Nac. Auton. Mex. 7 (1): 57–68.

Barlow, S.B., and R.E. Triemer. 1988. Alternate life history stages in *Amphidinium klebsii* (Dinophyceae, Pyrrophyta). Phycologia 27 (3): 413–420.

Barsanti, L., V. Evangelista, V. Passarelli, A.M. Frassanito, P. Colellai, and P. Gaultieri. 2009. Microspectrophotmetry as a method to identify kleptoplastids in the naked freshwater dinoflagellate *Gymnodinium acidotum*. J. Phycol. 45:1304–1309.

Baumeister, W. 1943. Die Dinoflagellaten der Kreise Pfarrkirchen und Eggenfelden (Gau Bayreuth). Arch. Protistenkd. 97:344-364.

Berdach, J.T. 1977. In situ preservation of the transverse flagellum of *Peridinium cinctum* (Dinophyceae) for scanning electron microscopy. J. Phycol. 13:243–251. [Note: actually *P. volzii*.]

Berg, K., and G. Nygaard. 1929. Studies on the plankton in the lake of Frederikborg Castle. Mém. Acad. R. Sci. Lett. Danemark 9 Série, 1:227–316, T1–6.

Berman-Frank, I., and J. Erez. 1996. Inorganic carbon pools in the bloom-forming dinoflagellate *Peridinium gatunense*. Limnol. Oceanogr. 41 (8): 1780–1789.

Bicudo, C.E.M., and B.V. Skvortzov. 1968. Contribution to the knowledge of Brazilian Dinophyceae immobile genera. An. Soc. Bot. Brasil, An. 19 Congr. Soc. Bot. Brasil, Fortaleza 31–39, Figs. 1–24.

Bicudo, C.E.M., and B.V. Skvortzov. 1970. Contribution to the knowledge of Brazilian Dinophyceae—free-living unarmored genera. Rickia 5:5–21, Figs. 1–22.

Bint, A.N. 1983. *Umbodinium crustov* gen. et sp. nov., a peridinioid dinoflagellate with two intercalaries from the Albian of Kansas. Palynology 7:171–182.

Bint, A.N. 1986. Fossil Ceratiaceae: a restudy and new taxa from the mid-Cretaceous of the western interior, U.S.A. Palynology 10:135–180.

Boltovskoy, A. 1975. Estructura y estereoultraestructura tecal de dinoflagelados. II. *Peridinium cinctum* (Müller) Ehrenberg. Physis (B) 34:73–84.

Boltovskoy, A. 1989. Thecal morphology of the dinoflagellate *Peridinium gutwinskii*. Nova Hedwigia 49 (3–4): 369–380.

Boltovskoy, A. 1999. The genus *Glochidinium* gen. nov., with two species: *G. penardiforme* com. nov. and *G. platygaster* sp. nov. (Peridiniaceae). Grana 38:98–107.

Botes, L., B. Price, M. Waldron, and G.C. Pitcher. 2002. A simple and rapid scanning electron microscope preparative technique for delicate "Gymnodinioid" dinoflagellates. Microsc. Res. Tech. 59:128–130.

Bourrelly, P. 1966. Quelques algues d'eau douce du Canada. Int. Rev. Gesamt. Hydrobiol. 51:45–126.

Bourrelly, P. 1968a. Notes sur les Péridiniens d'eau douce. Protistologica 4:5–16.

Bourrelly, P. 1968b. Note sur *Peridiniopsis borgei* Lemm. Phykos 7:1–2.

Bourrelly, P. 1970. *Les Algues d'Eau Douce*. Initiation à la Systématique. Tome III Les algues bleues et rouges Les Eugléniens, Peridiniens et Cryptomonadines. Éditions N. Boubée & Cie, Paris.

Bourrelly, P., and A. Couté. 1976. Observations en microscopie électronique à balayage des *Ceratium* d'eau douce (Dinophycées). Phycologia 15:329–338.

Bourrelly, P., and E. Manguin. 1952. Algues d'eau douce de la Guadeloupe et dépendances. Recueillies par la Mission P. Allorge en 1936. Société D'Edition D'Enseignment Supérieur, Paris.

Bovee, E. 1960. Protozoa of the Mountain Lake region, Giles Co, Virginia. J. Protozool. 7:352–361.

Brehm, V. 1907. Beiträge zur faunistischen Durchforschung der Seen Nordtirols. Naturw.-med. Verein Innsbruck. 31:97–120.

Brunel, J., and M. Poulin. 1992. Inventaire des algues d'eau douce de deux territoires protégés des Laurentides (Québec), de 1951 à 1966. Provancheria 26:1–50.

Bruno, S.F., and J.J.A. McLaughlin. 1977. The nutrition of the freshwater dinoflagellate *Ceratium hirundinella*. J. Protozool. 24:548–553.

Buckland-Nicks, J., and T.E. Reimchen. 1995. A novel association between an endemic stickleback and a parasitic dinoflagellate. 3. Details of the life cycle. Arch. Protistenkd. 145:165–175.

Buckland-Nicks, J., T.E. Reimchen, and M.F.J.R Taylor. 1990. A novel association bewteen an endemic stickleback and a parasitic dinoflagellate. 2. Morphology and life cycle. J. Phycol. 26:539–548.

Buckland-Nicks, J., T.E. Reimchen, and D.J. Garbary. 1997. *Haidadinium ichthyophilum* gen. nov. et sp. nov. (Phytodiniales, Dinophyceae), a freshwater ectoparasite on stickleback (*Gasterosteus aculeatus*) from the Queen Charlotte Islands, Canada. Can. J. Bot. 75:1936–1940.

Bursa, A. 1958. The freshwater dinoflagellate *Woloszynskia limnetica* n. sp. membrane and protoplasmic structures. J. Protozool. 5 (4): 299–304.

Bursa, A.S. 1969. *Actiniscus canadensis* n. sp., *A. pentasterias* Ehrenberg v. *arcticus* n. var., *Pseudoactiniscus apentasterias* n. gen., n. sp., marine relicts in Canadian arctic lakes. J. Protozool. 16:411–418.

Bursa, A.S. 1970. *Dinamoebidium coloradense* spec. nov. and *Katodinium auratum* spec. nov. in Como Creek, Boulder County, Colorado. Arct. Alp. Res. 2:145–151.

Calado, A.J. 2011. On the identity of the freshwater dinoflagellate *Glenodinium edax,* with a discussion on the genera *Tyrannodinium* and *Katodinium,* and the description of *Opisthoaulax* gen. nov. Phycologia 50 (6): 641–649.

Calado, A.J., and J.M. Huisman. 2010. Commentary: Gómez, F., Moreira, D., and López-García, P. (2010). *Neoceratium* gen. nov., a new genus for all marine species currently assigned to *Ceratium* (Dinophyceae). Protist 161:35–54, 517–519.

Calado, A.J., and J. Larsen. 1997. On the identity of the type species of the genus *Ceratium* Schrank (Dinophyceae), with notes on *C. hirundinella*. Phycologia 36 (6): 500–505.

Calado, A.J., and Ø. Moestrup. 2002. Ultrastructural study of the type species of *Peridiniopsis, Peridiniopsis borgei* (Dinophyceae), with special reference to the peduncle and flagellar apparatus. Phycologia 41 (6): 567–584.

Calado, A.J., and Ø. Moestrup. 2005. On the freshwater dinoflagellates presently included in the genus *Amphidinium,* with a description of *Prosoaulax* gen. nov. Phycologia 44:112–119.

Calado, A.J., S.C. Craveiro, and Ø. Moestrup. 1998. Taxonomy and ultrastructure of a freshwater heterotrophic *Amphidinium* (Dinophyceae) that feeds on unicellular protists. J. Phycol. 34:536–554.

Calado, A.J., G. Hansen, and Ø. Moestrup. 1999. Architecture of the flagellar apparatus and related structures in the type species of *Peridinium, Peridinium cinctum* (Dinophyceae). Eur. J. Phycol. 34:179–191.

Calado, A.J., S. C. Craveiro, N. Daugbjerg, and Ø. Moestrup. 2006. Ultrastructure and LSU rDNA-based phylogeny of *Esoptrodinium gemma* (Dinophyceae), with notes on feeding behavior and the description of the flagellar base area of a planozygote. J. Phycol. 42:434–452.

Calado, A.J., S.C. Craveiro, N. Daugbjerg, and Ø. Moestrup. 2009. Description of *Tyrannodinium* gen. nov., a freshwater dinoflagellate closely related to the marine *Pfiesteria*-like species. J. Phycol. 45:1195–1205.

Cameron, E. 1964. Algae of southern Arizona. Part II Algal flora (exclusive of blue-green algae). Rev. Algol. 7:151–175+.

Canion, A.K., and C. Ochs. 2005. The population dynamics of freshwater armored dinoflagellates in a small lake in Mississippi. J. Freshw. Ecol. 20:617–626.

Cantoral-Uriza, E.A., and G. Montejano-Zurita. 1993. Algae from El Sulto, San Luis Potosi, Mexico: an example of floristic studies in changing environment. Bol. Soc. Bot. Mex. 53:3–20.

Carefoot, J.R. 1968. Culture and heterotrophy of the freshwater dinoflagellate *Peridinium cinctum* fa. *ovoplanum* Lindemann. J. Phycol. 4:129–131.

Carrick, H.J. 2005. An under-appreciated component of biodiversity in plankton communities: the role of protozoa in Lake Michigan (a case study). Hydrobiologia 551:17–32.

Carty, S. 1986. The taxonomy and systematics of freshwater armored dinoflagellates. PhD dissertation. Texas A&M Univ., College Station.

Carty, S. 1989. *Thompsodinium* and two species of *Peridiniopsis* (Dinophyceae): taxonomic notes based on scanning electron micrographs. Trans. Am. Microsc. Soc. 108:64–73.

Carty, S. 1993. Contribution to the dinoflagellate flora of Ohio. Ohio J. Sci. 93: 140–146.

Carty, S. 2003. Dinoflagellates. In J.D. Wehr and R.G. Sheath, eds., Freshwater algae of North America: ecology and classification. Academic, 685–714.

Carty, S. 2007. New records for freshwater dinoflagellates in Alaska. Ohio J. Sci. 107 (1): A26.

Carty, S. 2008. *Parvodinium* gen. nov. for the Umbonatum Group of *Peridinium* (Dinophyceae). Ohio J. Sci. 108 (5): 103–107.

Carty, S. 2009. New records for freshwater dinoflagellates in New England, New Brunswick and Nova Scotia. Ohio J. Sci. 109:A5.

Carty, S., and E.R. Cox. 1985. Observations on *Lophodinium polylophum* (Dinophyceae). J. Phycol. 21:396–401.

Carty, S., and E.R. Cox. 1986. *Kansodinium* gen. nov. and *Durinskia* gen. nov.: two genera of freshwater dinoflagellates (Pyrrhophyta). Phycologia 25:197–204.

Carty, S., and J.D. Hall. 2002. Desmids and dinoflagellates of Ecuador. Abstract. J. Phycol. 38 (s1).

Carty, S., and D.E. Wujek. 2003. A new species of *Peridinium* and new records of dinoflagellates and silica-scaled chrysophytes from Belize. Caribb. J. Sci. 39: 136–139.

Chapman, A.D., and L.A. Pfiester. 1995. The effects of temperature, irradiance, and nitrogen on the encystment and growth of the freshwater dinoflagellates *Peridinium cinctum* and *P. willei* in culture (Dinophyceae). J. Phycol. 31:355–359.

Chesnick, J. M., and E.R. Cox. 1985. Thecal plate tabulation and variation in *Peridinium balticum* (Pyrrhophyta: Peridiniales). Trans. Am. Microsc. Soc. 104 (4): 387–394.

Chesnick, J. M., and E.R. Cox. 1987. Synchronized sexuality of an algal symbiont and its dinoflagellate host, *Peridinium balticum* (Levander) Lemmermann. BioSystems 21:69–78.

Chesnick, J. M., and E.R. Cox. 1989. Fertilization and zygote development in the binucleate dinoflagellate *Peridinium balticum* (Pyrrhophyta). Am. J. Bot.76: 1060–1072.

Chesnick, J. M., Morden, C.W., and Schmieg A.M. 1996. Identity of the endosymbiont of *Peridinim foliaceum* (Pyrrhophyta): analysis of the rbcLS operon. J. Phycol. 32:850–857.

Chesnick, J. M., W.H.C.F. Kooistra, U. Wellbrock, and L.K. Medlin. 1997. Ribosomal RNA analysis indicates a benthic pennate diatom ancestry for the endosymbionts of the dinoflagellates *Peridinium foliaceum* and *Peridinium balticum* (Pyrrhophyta). J. Eukaryot. Microbiol. 44 (4): 314–320.

Chodat, R. 1923. Algues de la région du Grand Saint Bernard. Bull. Soc. Bot. Genève 14:33–48.

Christen, Von H.R. 1961. Über die Gattung *Katodinium* Fott (= *Massartia* Conrad). Schweiz. Z. Hydrol. 23:309–341.

Christensen, T. 1978. Annotations to a textbook of phycology. Bot. Tidskrift 73 (2): 65–70.

Claparède, E., and J. Lachmann. 1858. Études sur les infusoires et les rhizopodes. Mém. Inst. Nat. Génèvois 5: 1–260.

Claparède, E., and J. Lachmann. 1859. Études sur les infusoires et les rhizopodes. Mém. Inst. Nat. Génèvois 6: 261–482, Pl. 14–24.

Clark, H.W. 1908. The holophytic plankton of Lakes Atitlan and Amatitlan, Guatemala. Proc. Biol. Soc. Wash. 11:91–106.

Coats, D.W., S. Kim, T.R. Bachvaroff, S.M. Handy, and C.F. Delwiche. 2010. *Tintinnophagus acutus* n. g., n. sp. (Phylum Dinoflagellata), an ectoparasite of the ciliate *Tintinnopsis cylindrical* Daday 1887, and its relationship to *Duboscquodinium collini* Grassé 1952. J. Eukaryot. Microbiol. 57 (6): 468–482.

Cohen, R.R.H. 1985. Physical processes and the ecology of a winter dinoflagellate bloom of *Katodinium rotundatum*. Mar. Ecol.—Prog. Ser. 26:135–144.

Cole, G.A. 1957. Studies on a Kentucky Knobs lake. III. Some qualitative aspects of the net plankton. Trans. Ky. Acad. Sci. 18:88–101.

Colt, L.C. 1999. A guide to the algae of New England. A comprehensive guide to the algae of the New England region as reported in the literature. PIPress, Walpole, MA, USA [in two parts: algae A–N in part 1, O–Z and literature in part 2].

Conn, H.W. 1905. The protozoa of the fresh waters of Connecticut. Conn. Geol. Nat. Hist. Survey Bull. No. 2.

Conrad, W. 1926. Researches sur les Flagellates de nos eaux saumâtres I. Partie: Dinoflagellates. Arch. Protistenkd. 55:63–100 + Plates 9 and 10.

Couté, A., and G. Tell. 1990. Quelques *Peridinium* Ehrbg. (Algae, Pyrrhophyta) d'eau douce étudiés au microscope électronique a balayage. Cryptogamie: Algol. 11 (3): 203–218.

Craveiro, S.C., A.J. Calado, N. Daugbjerg, and Ø. Moestrup. 2009. Ultrastructure and LSU rDNA-based revision of Peridinium group Palatinum (Dinophyceae) with the description of *Palatinus* gen. nov. J. Phycol. 45:1175–1194.

Craveiro, S.C., A.J. Calado, N. Daugbjerg, G. Hansen, and Ø. Moestrup. 2011. Ultrastructure and LSU rDNA-based phylogeny of *Peridinium lomnickii* and description of *Chimonodinium* gen. nov. (Dinophyceae). Protist 162: 590–615.

Craveiro, S.C., Ø. Moestrup, N. Daugbjerg, and A.J. Calado. 2010. Ultrastructure and large subunit rDNA-based phylogeny of *Sphaerodinium cracoviense*, an unusual freshwater dinoflagellate with a novel type of eyespot. J. Eukaryot. Microbiol. 57 (6): 568–585.

Crawford, R.M., J.D. Dodge, and C.M. Happey. 1970. The dinoflagellate genus *Woloszynskia*. I. Fine structure and ecology of *W. tenuissimum* from Abbot's Pool, Somerset. Nova Hedwigia 19:825–840.

Croasdale, H.T. 1948. The fresh and brackish water algae of Penikese Island. Rhodora 50:270–280.

Croome, R.L., and P.A. Tyler. 1987. *Prorocentrum playfairi* and *Prorocentrum foveolata,* two new dinoflagellates from Australian freshwaters. Br. Phycol. J. 22: 67–75.

CSLAP. 2003. New York citizens statewide lake assessment program. 2002 Interpretive summary. NY Federation of Lake Associations. NYS Department of Environmental Convervation.

Da, K.P., F. Zongo, G. Mascarell, and A. Couté. 2004. *Bagredinium,* un nouveau genre de Péridiniales (Dinophyta) d'eau douce de l'Afrique de l'Ouest. Algol. Stud. 111:45–61.

Daday, E. 1905. Untersuchungen über Süswasser-Mikrofauna Paraguays. Zoologica 18 (44): 1–374.

Daday, E. 1907. Plancton-Tiere aus dem Victoria Nyanza. Sammelausbeute von A. Borgert 1904–1905. Zool. Jb. Abt. Syst. 25:245–262.

Daily, W.A., and E.F. Miner. 1953. The phytoplankton of Lake Wawasee, Kosciusko County, Indiana. Butler Univ. Bot. Stud. 11:91–99.

Danysz, J. 1887. Contribution à l'étude de l'evolution des Péridiniens d'eau douce. C.R. Acad. Sci. Paris 105:238 [refers to *Gymnodinium musei,* previously described in Pouchet 1887].

Daugbjerg, N., G. Hansen, J. Larsen, and Ø. Moestrup. 2000. Phylogeny of some of the major genera of dinoflagellates based on ultrastructure and partial LSU rRNA sequence data, including the erection of three new genera of unarmoured dinoflagellates. Phycologia 39 (4): 302–317.

Davis, C.C. 1972. Plankton succession in a Newfoundland lake. Int. Rev. Gesamt. Hydrobiol. Hydrogr. 57:267–395.

Davis, C.C. 1973. A seasonal quantitative study of the plankton of Bauline Long Pond, a Newfoundland lake. Nat. Can. 100:85–105.

Davis, R.B., J.H. Bailey, M. Scott, G. Hunt, and S.A. Norton. 1978. Descriptive and comparative studies of Maine lakes. Tech. Bull. 88 Life Sci. Agri. Exp. Stat. Orono, ME.

Dawes, C., and J. Jewett-Smith. 1985. Algal flora of an acid marsh in west-central Florida. Trans. Am. Microsc. Soc. 104:188–193.

Dickman, M.D. 1968. The relation of freshwater plankton productivity to species composition during induced succession. PhD thesis, Dept. Zoology, Univ. B.C., Vancouver.

Diesing, K.M. 1866. Revision der Prothelminthen, Abtheilung: Mastigophoren. Sitzungsberichte der Mathematisch-naturwissenschaftliche Klasse, 52 (8): 287–401. Akademie der Wissenschaften zu Wein.

Dillard, G. 1967. The freshwater algae of South Carolina, I. J. Elisha Mitchell Sci. Soc. 83:128–132.

Dillard, G.E. 1974. An annotated catalog of Kentucky algae. Western Kentucky Univ., Ogden College of Science and Technology.

Dillard, G.E. 2007. Freshwater algae of the southeastern United States. Part 8. Chrysophyceae, Xanthophyceae, Raphidophyceae, Cryptophyceae, and Dinophyceae. Bibliotheca Phycologia Band 112. J. Cramer, Berlin.

Dillard, G.E., and S.B. Crider. 1970. Kentucky algae. I. Trans. Ky. Acad. Sci. 31: 66–72.

Dillard, G.E, S.P. Moore, and L.S. Garrett. 1976. Kentucky algae. II. Trans. Ky. Acad. Sci. 37:20–25.

Dodge, J.D. 1963. The nucleus and nuclear division in the Dinophyceae. Arch. Protistenkd. 106:442–453.

Dodge, J.D. 1969. A review of the fine structure of algal eyespots. Br. Phycol. J. 4:199–210.

Dodge, J.D. 1972. The ultrastructure of the dinoflagellate pusule: a unique osmo-regulatory organelle. Protoplasma 75:285–302.

Dodge, J.D. 1975. A survey of chloroplast ultrastructure in the Dinophyceae. Phycologia 14:253–263.

Dodge, J.D. 1983. Ornamentation of thecal plates in *Protoperidinium* (Dinophyceae) as seen by scanning electron microscopy. J. Plankton Res. 5:119–127.

Dodge, J.D., and R.M. Crawford. 1970. The morphology and fine structure of *Ceratium hirundinella* (Dinophyceae). J. Phycol. 6:137–149.

Dodge, J.D., and H.B. Hermes. 1981. A scanning electron microscopical study of the apical pores of marine dinoflagellates (Dinophyceae). Phycologia 20:424–430.

Dodge, J.D., and K. Vickerman. 1980. Mitosis and meiosis: nuclear division mechanisms. In G.W. Gooday, D. Lloyd, and A.P. Trinci, eds., The eucaryotic microbial cell. Society for General Microbiology. Symposium 30. Cambridge Univ. Press.

Dujardin, F. 1841. Histoire naturelle des Zoophytes. Infusoires, comprenant la physiologie et la classification de ces animaux et la manière de les étudier à l'aide du microscope. Librairie Encyclopedique de Roret, Paris.

Durrell, L.W., and C. Norton. 1960. Phytoplankton of Lakes of Grand Mesa, Colorado. Trans. Am. Microsc. Soc. 79:91–97.

Duthie, H.C., and R. Socha. 1976. A checklist of the freshwater algae of Ontario, exclusive of the Great Lakes. Nat. Can. 103:83–109.

Duthie, H.C., M.L. Ostrofsky, and D.J. Brown. 1976. Freshwater algae from western Labrador IV. Chrysophyta, Xanthophyta, Pyrrhophyta, Cryptophyta. Nova Hedwigia 27:909–917.

Eaton, G. 1980. Nomenclature and homology in Peridinialean dinoflagellate plate patterns. Palaeontology 23 (3): 667–688.

Eddy, S. 1930. The fresh-water armored or thecate dinoflagellates. Trans. Am. Microsc. Soc. 49:277–321.

Ehrenberg, C.G. 1830. Beitrage zur Kenntnis der Organisation der Infusorien and ihrer geographischen Verbreitung, besonders in Sibirien. Abh. Königl. Akad. Wiss. Berlin.

Ehrenberg, C.G. 1834 (separate), 1835. Dritter Beitrag zur Erkentniss grosser Organisation in der kleinsten Raumes. Abhandlungen der Königlichen Akademie der Wissenschaften zu Berlin 1833, Physikalische Klasse:145–336, Pl. 1–11.

Ehrenberg, C.G. 1838. Die Infusionsthierchen als volkommene Organismen. Leopold Voss, Berlin und Leipzig.

Ehrenberg, C.G. 1841. Über noch jetzt zahlreich lebende Thierarten der Kreidebildung und den Organismus der Polythalamien. Königlich Preussische Akademie der Wissenschaften zu Berlin, Bericht über die zur Bekanntmachung geeigneten Verhandlungen 1839:81–174, Pl. 1–4. [*Dictyocha* subgenus *Actiniscus*]

Ehrenberg, C.G. 1843. Über die Verbreitung des jetzt wirkenden kleinsten organisischen Lebens in Asien, Australien und Afrika und über die vorherrschende Bildung auch des Oolithkalkes der Juraformation aus kleinen polythalamischen Thieren. Königlich Preussische Akademie der Wissenschaften zu Berlin, Bericht über die zur Bekanntmachung geeigneten Verhandlungen, 1844:100–106. [*Actiniscus*]

Entz, G. 1904, June. Beiträge zur Kenntnis des Planktons des Balatonsees Resultate d. Wiss. Erforsch. Des Balatonsees, Bd 2, I Teil, Anhang, Budapest. 1–37. With 17 text figures and 9 plates [transfers *Peridinium apiculatum* to *Gonyaulax*]

Entz, G. 1925. Über Cysten und Encystierung der Süßwasserceratien. Arch. Protistenkd. 51:131–182.

Entz, G. 1927. Beiträge zur Kenntnis der Peridineen. II. Resp. VII Studien an Süßwasser-Ceratien. (Morphologie, Variation, Biologie.). Arch. Protistenkd. 58:344–440.

Evitt, W.R. 1985. Sporopollenin dinoflagellate cysts: their morphology and interpretation. AASP Foundation.

Evitt, W.R., and D. Wall. 1968. Dinoflagellates studies. IV Theca and cyst of recent freshwater *Peridinium limbatum* (Stokes) Lemmermann. Stanford Univ. Publ. Geol. Sci. 12 (2): 1–15, Pl. 1–4.

Farmer, M.A., and K.R. Roberts. 1990. Comparative analysis of the dinoflagellate flagellar apparatus 4. *Gymnodinium acidotum.* J. Phycol. 26:122–131.

Fawcett, R.C. and M.W. Parrow, 2012. Cytological and phylogenetic diversity in freshwater *Esoptrodinium/Bernardinium* species (Dinophyceae). J. Phycol. 48:793–807.

Fenchel, T. 2001. How dinoflagellates swim. Protist 152:329–338.

Fensome, R.A., F.J.R. Taylor, G. Norris, W.A.S. Sarjeant, D.I. Wharton, and G.L. Williams. 1993. A classification of living and fossil dinoflagellates. Micropaleontology Spec. Publ. 7.

Fields, S.D., and R.G. Rhodes. 1991. Ingestion and retention of *Chroomonas* spp. (Cryptophyceae) by *Gymnodinium acidotum* (Dinophyceae). J. Phycol. 27:525–529.

Figueroa-Torres, G., and J.L. Moreno-Ruiz. 2003. Dinoflagelados dulceacuícolas de México. In T. Barreiro-Güemes, E.M. del Castillo, M. Signoret-Poillon, and G. Figueroa-Torres, eds., Planctología Mexicana. Sociedad Mexicana de Planctología, Baja California Sur, México, 85–102.

Findlay, D.L. 2003. Response of phytoplankton communities to acidification and recovery in Killarney Park and the Experimental Lakes Area, Ontario. Ambio 32 (3): 190–195.

Foerster, J.W. 1971. The ecology of an elfin forest in Puerto Rico, 14. The algae of Pico del Oeste. J. Arnold Arboretum 52: 86–109.

Forest, H.S. 1954a. Handbook of Algae with special reference to Tennessee and the southeastern United States. Univ. Tenn. Press

Forest, H.S. 1954b. Checklist of algae in the vicinity of Mountain Lake Biological Station—Virginia. Castanea 19:88–104.

Fott, B. 1938. Eine neue *Gymnodinium-* und *Massartia-*Art. Studia Botanica Čechoslovaca 1:100–104.

Fott, B. 1957. Taxonomie drobnohledné flory našich vod. Preslia 29:278–319.

Fott, B. 1960. Taxonomische Übertragungen und Namensäanderungen unter den Algen. Preslia 32:142–154.

Fott, J., M. Blažo, E. Stuchlík, and O. Strunecký. 1999. Phytoplankton in three Tatra Mountain lakes of different acidification status. J. Limnol. 58 (2): 107–116.

Frey, L.C., and E.F. Stoermer. 1980. Dinoflagellate phagotrophy in the upper Great Lakes. Trans. Am. Microsc. Soc. 99:439–444.

Geitler, L. 1928a. Zwei neue Dinophyceenarten. Arch. Protistenkd. 61:1–8. [*Cystodinium iners* original description]

Geitler, L. 1928b. Neue Gattungen und Arten von Dinophyceen, Heterokonten und Chrysophyceen. Arch. Protistenkd. 63:67–83. [*Raciborskia inermis*]

Geitler, L. 1943. Koleoniebidung und Beeinflussung der Unterlage bei zwei Dinococcalen (*Raciborskia oedogonii* und *R. inermis* n. sp.). Beih. Bot. Centralblatt 62: A 160–174.

Gerrath, J.F., and K.H. Nicholls. 1974. A red snow in Ontario caused by the dinoflagellate, *Gymnodinium pascheri*. Can. J. Bot. 52:683–685.

Gilpin, C. 2012. Diel temperature and dissolved oxygen patterns in sites with and without planktonic life stage of *Thompsodinium intermedium* in Comal Springs, TX. MS thesis, Texas A&M Univ.

Goldstein, A.K., and J.J. Manzi. 1976. Additions to the freshwater algae of South Carolina. J. Elisha Mitchell Sci. Soc. 92:9–13.

Gómez, F. 2010. A genus name for the marine species of *Ceratium*. Reply to commentary by A. Calado and J.M. Huisman on Gómez, F., Moreira, D., and López-García, P. 2010. *Neoceratium* gen. nov., a new genus for all marine species currently assigned to *Ceratium* (Dinophyceae). Protist 161:35–54, 520–522.

Gómez, F., D. Moreira, and P. López-García. 2010. *Neoceratium* gen. nov., a new genus for all marine species currently assigned to *Ceratium* (Dinophyceae). Protist 161:35–54.

Graham, J.M., A.D Kent, G.H. Lauster, A.C. Yannarell, L.E. Graham, and E.W. Triplett. 2004. Seasonal dynamics of phytoplankton and planktonic protozoan communities in a northern temperate humic lake: diversity in a dinoflagellate dominated system. Microb. Ecol. 48:528–540.

Grigorszky, I., L. Krienitz, J. Padisak, G. Borics, and G. Vasas. 2003. Redefinition of *Peridinium lomnickii* Woloszynska (Dinophyta) by scanning electronmicroscopical survey. Hydrobiologia 502:349–355.

Gruendling, G.K., and A.C. Mathieson. 1969. Phytoplankton flora of Newfound and Winnisquam Lakes, New Hampshire. Rhodora 71:444–477.

Gustafson, A.H. 1942. Notes on some fresh-water algae from New England. Rhodora 44:64–69.

Haberyan, K.A., G. Umaña, C. Collado, and S.P. Horn. 1995. Observations on the plankton of some Costa Rican lakes. Hydrobiologia 312:75–85.

Hallegraeff, G.M. 2010. Ocean climate change, phytoplankton community responses, and harmful algal blooms: a formidable predictive challenge. J. Phycol. 46: 220–235.

Hamlaoui, S., A. Couté, G. Lacroix, and F. Lescher-Moutoué. 1998. Nutrient and fish effects on the morphology of the dinoflagellate *Ceratium hirundinella*. C.R. Acad. Sci., Ser. III Sci. Vie/Life Sci. 321:39–45.

Hammer, U.T., J. Shamess, and R.C. Haynes. 1983. The distribution and abundance of algae in saline lakes of Saskatchewan, Canada. Hydrobiologia 105:1–26.

Hanisak, M.D. 1973. An ecological survey of the phytoplankton of the Pettaquamsett River, Rhode Island. MS thesis, Univ. R.I., Kingston.

Hansen, G. 1993. Light- and electron microscopical observations on the dinoflagellates *Actiniscus pentasterias* (Dinophyceae). J. Phycol. 29:486–499.

Hansen, G. 1995. Analysis of the thecal plate pattern in the dinoflagellate *Heterocapsa rotundata* (Lohmann) comb. nov. (= *Katodinium rotundatum* (Lohmann) Loeblich). Phycologia 34 (2): 166–170.

Hansen, G., and N. Daugbjerg. 2004. Ultrastructure of *Gyrodinium spirale,* the type species of *Gyrodinium* (Dinophyceae), incuding a phylogeny of *G. dominans, G. rubrum* and *G. spirale* deduced from partial LSU rDNA sequences. Protist 155:271–294.

Hansen, G., Ø. Moestrup, and K.R. Roberts. 2000. Light and electron microscopical observations on the type species of *Gymnodinium, G. fuscum* (Dinophyceae). Phycologia 39 (5): 365–376.

Hansen, K. 1967. The general limnology of Arctic lakes as illustrated by examples from Greenland. Medd. Grønland 178:51–60.

Happach-Kasan, C. 1982. Beobachtungen zum Bau der Theka von *Ceratium cornutum* (Ehrenb,) Clap. et Lachm. (Dinophyta). Arch. Protistenkd. 125:181–207.

Hargraves, P.E., and R.M. Víquez. 1981. Dinoflagellate abundance in the Laguna Botos, Poás Volcano, Costa Rica. Rev. Biol. Trop. 29:257–264.

Harris, T.M. 1940. A contribution to the knowledge of the British freshwater dinoflagellata. Proc. Linn. Soc. Lond. 152nd session, 1939–1940, Part 1: 4–33.

Hashimoto, Y., T. Okaichi, L.D. Dang, and T. Noguchi. 1968. Glenodine, an ichthyotoxic substance produced by a dinoflagellate, *Peridinium polonicum*. Bull. Jap. Soc. Sci. Fish. 34 (6): 528–534.

Hassall, A.H. 1845. A history of the British freshwater algae including descriptions of the Desmideae and Diatomaceae. S. Highly and H. Balliere, London.

Hayhome, B.A., and L.A. Pfiester. 1983. Electrophoretic analysis of soluble enzymes in five freshwater dinoflagellate species. Am. J. Bot. 70 (8): 1165–1172.

Hayhome, B.A., D.J. Whitten, K.R. Harkins, and L.A. Pfiester. 1987. Intraspecific variation in the dinoflagellate *Peridinium volzii*. J. Phycol. 23:573–580.

Hecky, R.E., H.J. Kling, and G.J. Brunskill. 1986. Seasonality of phytoplankton in relation to silicon cycling and interstitial water circulation in large, shallow lakes of central Canada. Hydrobiologia 138:117–126.

Herbst, R.P., and R.T. Hartman. 1981. Phytoplankton distribution of a duckweed covered pond. J. Freshw. Ecol. 1:97–112.

Herdman, E.C. 1924. Notes on dinoflagellates and other organisms causing discolourations of the sand at Port Erin. IV. Proc. Trans. Liverpool Biol. Soc. 38:75–84.

Hern, S.C., V.W. Lambou, F.A. Morris, M.K. Morris, W.D. Taylor, and L.R. Williams. 1979a. Distribution of phytoplankton in Oklahoma lakes. EPA-600/3-79-068.

Hern, S.C., V.W. Lambou, F.A. Morris, M.K. Morris, W.D. Taylor, and L.R. Williams. 1979b. Distribution of phytoplankton in Montana lakes. EPA-600/3-79-116.

Hern, S.C., V.W. Lambou, F.A. Morris, M.K. Morris, W.D. Taylor, and L.R. Williams. 1979c. Distribution of phytoplankton in South Dakota lakes. EPA-600/3-79-069.

Herrgesell, P.L., T.H. Sibley, and A.W. Knight. 1976. Some observations on dinoflagellate population density during a bloom in a California reservoir. Limnol. Oceanogr. 21:619–624.

Hickel, B. 1988. Sexual reproduction and life cycle of *Ceratium furcoides* (Dinophyceae) in situ in the lake Plußsee (F.R.G.). Hydrobiologia 161:49–54.

Hickel, B., and U. Pollinger. 1988. Identification of the bloom-forming *Peridinium* from Lake Kinneret (Israel) as *P. gatunense* (Dinophyceae). Br. Phycol. J. 23: 115–119.

Highfill, J.F., and L.A. Pfiester. 1992a. The sexual and asexual life cycle of *Glenodiniopsis steinii* (Dinophyceae). Am. J. Bot. 79:899–903.

Highfill, J.F., and L.A. Pfiester, L.A. 1992b. The ultrastructure of *Glenodiniopsis steinii* (Dinophyceae). Am. J. Bot. 79:1162–1170.

Hilgert, J.W. 1976. Flagellated algae of Flathead Valley. Proc. Mont. Acad. Sci. 36:86–97.

Hilgert, J.W., F.A. Morris, M.K. Morris, W.D. Taylor, L.R. Williams, S.C. Hern, and V.W. Lambou. 1978a. Distribution of phytoplankton in Arkansas lakes. EPA-600/3-78-101.

Hilgert, J.W., V.W. Lambou, F.A. Morris, M.K. Morris, L.R. Williams, W.D. Taylor, F.A. Hiatt, and S.C. Hern. 1978b. Distribution of phytoplankton in Ohio lakes. EPA-600/3-78-015.

Hill, D.R. 1980. Phytoplankton ecology and trophic state analysis of Radnor Lake, Radnor Lake Natural Area, Nashville, Tennessee. PhD thesis, N.C. State Univ.

Hilliard, D.K. 1959. Notes on the phytoplankton of Karluk Lake, Kodiak Island, Alaska. Can. Field-Nat. 73:135–143.

Hirabayashi, K., K. Yoshizawa, N. Yoshida, K. Ariizumi, and F. Kazama. 2007. Long-term dynamics of freshwater red tide in shallow lake in central Japan. Environ. Health Prev. Med. 12:33–39.

Hoham, R.W. 1966. The freshwater algae of Mission Wells and Tykeson Pond, Montana. MS thesis, Mich. State Univ., E. Lansing.

Holopainen, I.J. 1992. The effects of low pH on planktonic sommunities. Case history of a small forest pond in eastern Finland. Ann. Zool. Fenn. 28:95–103.

Holt, J.R., and L.A. PfiesterA. 1981. A survey of auxotrophy in five freshwater dinoflagellates (Pyrrhophyta). J. Phycol. 17:415–416.

Holt, J.R., and L.A. Pfiester. 1982. A technique for counting chromosomes of armored dinoflagellates, and chromosome numbers of six freshwater dinoflagellate species. Am. J. Bot. 69 (7): 1165–1168.

Hooper, F.F. 1947. Plankton collections from the Yukon and Mackenzie River systems. Trans. Am. Microsc. Soc. 66:74–84.

Horne, A., P. Javornicky, and C. Goldman. 1972. A freshwater "red tide" of Clear Lake, California. Limnol. Oceanogr. 16:684–689.

Huber, G., and F. Nipkow. 1922. Experimentelle Untersuchyungen uber die Entwicklung von *Ceratium hirundinella* O.F.M. Zeitschr. Bot. 14: 337–371.

Huber-Pestalozzi, G. 1950. Das phytoplankton des Süsswassers. In A. Thienemann, ed., Die Binnengewasser Bd. 16, Teil 3. E. Schweizerhart'sche Verlangsbuchhandlung, Stuttgart.

Hughes, E.O. 1947–1948. Freshwater algae of the marine provinces. Nova Scotia Inst. Sci. 22 (2): 1–64.

Huitfeldt-Kaas. 1900. Die limnetischen Peridineen in norwegischen Binnenseen. Skr. Vidensk.-Selsk. Christiania, Math.-Naturvidensk. Kl 1900:1–7 and one plate.

Huszar, V., C. Kruk, and N. Caraco. 2003. Steady-state assemblages of phytoplankton in four temperate lakes (NE U.S.A.). Hydrobiologia 502:97–109.

Hutchinson, G.E. 1967. A treatise on limnology. Vol. 2, Introduction to lake biology and the limnoplankton. John Wiley & Sons, NY.

Iltis, A., and A. Couté. 1984. Péridiniales (algae, Pyrrhophyta) de Bolivie. Rev. Hydrobiol. Trop. 17 (4): 279–286.

Irish, A.E. 1979. *Gymnodinium helveticum* Penard f. *achroum* Skuja a case of phagotrophy. Br. Phycol. J. 14:11–15.

Jacobs, D.L. 1946. A new parasitic dinoflagellate from fresh-water fish. Trans. Am. Microsc. Soc. 65:1–17.

Jacobs, J.E. 1968. A preliminary checklist of the freshwater algae in South Carolina. J. Elisha Mitchell Sci. Soc. 84:454–457.

Jacobs, J.E. 1971. A preliminary taxonomic survey of the freshwater algae of the Belle W. Baruch plantation in Georgetown County, South Carolina. J. Elisha Mitchell Sci. Soc. 87:26–30.

Jacobson, D.M., and D.M. Anderson. 1986. Thecate heterotrophic dinoflagellates: feeding behavior and mechanisms. J. Phycol. 22:249–258.

Janus, L.L., and H.C. Duthie. 1979. Phytoplankton composition and periodicity in a northeastern Quebec lake. Hydrobiologia 63:129–134.

Javornický, P. 1962. Two scarcely known genera of the class Dinophyceae: *Bernardinium* Chodat and *Crypthecodinium* Biecheler. Preslia 34:98–113 + Tables 7–11.

Javornický, P. 1965. Unarmored Dinoflagellata from two small Mazurian lakes. Phycologia 5:53–60.

Javornický, P. 1971. *Peridinium penardii* (Lemm.) Lemm. fo. *Californicum,* forma nova. J. Phycol. 7:303–306.

Javornický, P. 1997. *Bernardinium* Chodat (Dinophyceae), an athecate dinoflagellate with reverse, right-handed course of the cingulum and transverse flagellum, and *Esoptrodinium* genus novum, its mirror-symmetrical pendant. Algol. Stud. Arch. Hydrobiol. (Suppl.) 87:29–42.

Johansen, J.R., R. Lowe, S.R. Gomez, J.P. Kociolek, and S.A. Makosky. 2004. New algal species records for the Great Smoky Mountains National Park, U.S.A., with an annotated checklist of all reported algal species for the park. Algol. Stud. 111:17–44.

Johansen, J.R., R.L. Lowe, S. Carty, K. Fučiková, C.E. Olsen, M.H. Fitzpatrick, J.R. Ress, and P.C. Furey. 2007. New algal species records for the Great Smoky Mountains National Park, U.S.A., with an annotated checklist of all reported algal species for the park. Southeast. Nat. 6 (Spec. Issue 1): 99–134.

Johnson, M.G., M.F.P. Michalski, and A.E. Christie. 1968. Effects of acid mine wastes on phytoplankton in northern Ontario lakes. Ont. Water Resour. Comm. Res. Publ. 30:16–21.

Jörgensen, E. 1911. Die Ceratien. Eine kurze Monographie der Gattung *Ceratium*. Int. Rev. Gesamt. Hydrobiol., 4 Suppl.: 1–124.

Jørgensen, M.F., S. Murray, and N. Daugbjerg. 2004. *Amphidinium* revisted. I. Redefinition of *Amphidinium* (Dinophyceae) based on cladistic and molecular phylogenetic analyses. J. Phycol. 40:351–365.

Joyce, R.E. 1936. The algae of Vermont. MA thesis, Univ. Vt., Burlington.

Kalff, J., H.J. Kling, S.H. Holmgren, and H.E. Welch. 1975. Phytoplankton, phytoplankton growth and biomass cycles in an unpolluted and in a polluted polar lake. Verh. Int. Ver. Theor. Angew. Limnol. 19:487–495.

Kawai, H., and G. Kreimer. 2000. Sensory mechanisms. Phototaxis and light perception in algae. In B.S.C. Leadbeater and J.C. Green, eds. The flagellates. Unity, diversity and evolution. Taylor and Francis, London, 124–146.

Kelley, I., and L.A. Pfiester. 1989. Vegetative reproduction of the freshwater dinoflagellate *Gloeodinium montanum*. J. Phycol. 25:241–247.

Kelley, I., and L.A. Pfiester. 1990. Sexual reproduction of the freshwater dinoflagellate *Gloeodinium montanum*. J. Phycol. 26:167–173.

Killian, C. 1924. Le cycle évolutif du *Gloeodinium montanum* (Klebs). Arch. Protistenkd. 50:50–66.

Kim, E., L. Wilcox, L. Graham, and J. Graham. 2004. Genetically distinct populations of the dinoflagellate *Peridinium limbatum* in neighboring northern Wisconsin lakes. Microb. Ecol. 48:521–527.

Kimmel, B.L., and J.R. Holt. 1988. Nutrient availability and patterns of polymorphism in the freshwater dinoflagellate, *Ceratium hirundinella*. Arch. Hydrobiol. 113 (4): 577–592.

Kiselev, I.A. 1954. Pirofitovye vodorosli. Opredelitel' presnovodnych vodoroslej SSSR 6. Sov. Nauka, Moskva, Leningrad.

Klarer, D.M., C.E. Herdendorf, and R.C. Herdendorf. 2000. Catalogue of the algal flora and lower plants of Old Woman Creek Estuary, watershed, and adjacent water of Lake Erie. Tech. Rep. 13. Contrib. 4. Appendix A, Algal flora, A1–A43.

Klebs, G. 1884. Ein kleiner Beitrag zur Kenntnis der Peridineen. Bot. Zeit. 42:737–752 + plates

Klebs, G. 1912. Über Flagellaten- und Algen-ähnliche Peridineen. Verhandlungen Naturhistorisch- Medizinischen Vereines zu Heidelberg, N.F. 11:369–451, Taf 10.

Klemer, A., and J. Barko. 1991. Effects of mixing and silica enrichment on phytoplankton seasonal succession. Hydrobiologia 210:171–181.

Kling, H.J., and S.K. Holmgren. 1972. Species composition and seasonal distribution in the experimental lakes area, northwestern Ontario. Fish. Res. Board Can. Tech. Rep. 337:29–49.

Knappe, D.R.U., R.C. Belk, D.S. Briley, S.R. Gandy, N. Rastogi, A.H. Rike, H. Glasgow, E. Hannon, W.D. Frazier, P. Kohl, and S. Pugsley. 2004. Algae detection and removal strategies for drinking water treatment plants. American Water Works Association.

Kobraei, M.E., and D.S. White. 1996. Effects of 2,4-dichlorophenoxyacetic acid on Kentucky algae: simultaneous laboratory and field toxicity testings. Arch. Environ. Contam. Toxicol. 31:571–580.

Kofoid, C.A. 1907. The plates of *Ceratium* with a note on the unity of the genus. Zool. Anz. 32:177–183.

Kofoid, C.A. 1908. The plankton of the Illinois River, 1894–1899, with introductory notes upon the hydrogeography of the Illinois River and its basin. Part II. Constituent organisms and their seasonal distribution. Bull. Ill. State Lab. Nat. Hist. 3, Article 1.

Kofoid, C.A. 1909. On *Peridinium steini* Jörgensen, with a note on the nomenclature of the skeleton of the Peridinidae. Arch. Protistenkd. 16:25–47.

Kofoid, C.A., and J.R. Michener 1912. On the structure and relationships of *Dinosphaera palustris* -(Lemm.). Univ. Calif. Pub. Zool. 11 (2): 21–28.

Kofoid, C.A., and O. Swezy. 1921. The free-living unarmoured dinoflagellates. Mem. Univ. Calif. 5:1–562.

Kolesar, S.E., D.M. McKnight, and S.B. Waters. 2002. Late fall phytoplankton dynamics in three lakes, Rocky Mountain National Park. Hydrobiologia 472:249–263.

Koryak, M. 1978. The occurrence of *Peridinium inconspicuum* Lemmermann (Dinophyceae) in minerally acid waters of the upper Ohio River basin. Proc. Acad. Nat. Sci. Phila. 130:22–25.

Krakhmalny, A.F., M.A. Gololobova, and M.A. Krakhmalny. 2006. Morphology of *Peridinium gatunense* Nyg. (Dinophyta) from Lake El Padre (Mexico). Int. J. Algae 8 (3): 211–218.

Kützing, F.T. 1849. Species algarum. F.A. Brockhaus, Leipzig.

Lackey, J.B. 1936. Some fresh-water Protozoa with blue chromatophores. Biol. Bull. 71:492–497.

Laíz, O., I. Quintana, P. Blomqvist, A. Broberg, and A. Infante. 1993a. Limnology of Cuban reservoirs: I. Lebrije. Trop. Freshw. Biol. 3:371–396.

Laíz, O., I. Quintana, P. Blomqvist, A. Broberg, and A. Infante. 1993b. Limnology of Cuban reservoirs: II. Higuanojo. Acta Cient. Venez. 44:297–306.

Lambou, V.W., F.A. Morris, M.K. Morris, W.D. Taylor, L.R. Williams, and S.C. Hern. 1979a. Distribution of phytoplankton in New Mexico lakes. EPA-600/3-79-118.

Lambou, V.W., F.A. Morris, M.K. Morris, W.D. Taylor, L.R. Williams, and S.C. Hern. 1979b. Distribution of phytoplankton in Nevada lakes. EPA-600/3-79-117.

Lambou, V.W., F.A. Morris, M.K. Morris, W.D. Taylor, L.R. Williams, and S.C. Hern. 1979c. Distribution of phytoplankton in Louisiana lakes. EPA-600/3-79-064.

Langhans, V.H. 1925. Gemischte Populationen von *Ceratium hirundinella* (O.F.M.) Schrank und ihre Deutung. Arch. Protistenkd. 52:585–602.

Larson, G.L., C.D. McIntire, R.E. Truitt, W.J. Liss, R. Hoffman, E. Deimling, and G. Lomnicky. 1998. Phytoplankton assemblages in high-elevation lakes in the northern Cascade Mountains, Washington State USA. Arch. Hydrobiol. 142:71–93.

Lauterborn, R. 1894. Über die Winterfauna einiger Gewässer der Oberrheinebene. Biol. Zentralbl. 14:390–398. [*Gymnodinium tenuissimum* description only]

Lauterborn, R. 1896. Diagnosen neuer Protozoen aus dem Gebiete des Oberrheins. Zool. Anz. 19:14–18.

Lauterborn, R. 1899. Protozoan-Studien. IV. Theil. Flagellaten aus dem Gebiete der Oberrheins. Z. Wiss. Zool. 656:369–391, Pl. 17, 18. [*Gymno. tenuissima* with figure]

Lebour, M.V. 1922. Plymouth peridinians. J. Mar. Biol. Assoc. Plymouth New Ser. 12 (4): 795–812.

Lefèvre M. 1925. Contribution à la flore des Péridiniens de France. Rev. Algol. 2:327–342.

Lefèvre M. 1927. Sur les variations tabulaires chez les Péridiniens d'eau douce et leur notation.—Diagnoses d'espèces et de variétés nouvelles. Bull. Mus. Natl. Hist. Nat. 33:118–122.

Lefèvre M. 1928. Notules systématiques. I. *Peridinium morzinense* nom. nov. Ann. Protistol. 1:137

Lefèvre, M. 1932. Monographie des espèces d'eau douce du genre *Peridinium*. Arch. Bot. Mem. Caen 2:1–208.

Lemmermann, E. 1899. Ergebnisse einer Reise nach dem Pacific (H. Schauinsland 1896/97). Planktoalgen. Abh. Naturwiss. Ver. Bremen 16:313–398, Pl. 1–3.

Lemmermann, E. 1900a. Beiträge zur Kenntnis der Planktonalgen. VIII. Peridiniales aquae dulcis et submarinae. Hedwigia 39:115–121. [*Gl. penardi, P. balticum, P. limbatum*]

Lemmermann, E. 1900b. Beiträge zur Kenntnis der Planktonalgen. III. Neue Schwebalgen aus der Umgegend von Berlin. Ber. Dtsch. Bot. Ges. 18:24–32. [*P. marsonii* and *P. aciculiferum*]

Lemmermann, E. 1900c. Beiträge zur Kenntnis der Planktonalgen. X. Diagnosen neuer schwebalgen. Ber. Dtsch. Bot. Ges. 18:306–310. [*Peridinium/Tyranodinium berolinense*]

Lemmermann, E. 1904. Das plankton schwedischer Gewasser. Arkiv. Bot. 2:1–209.

Lemmermann, E. 1906. Über die von Herrn Dr. Walter Volz auf seiner weltreise gesammelten Süsswassalgen. Abh. Naturwiss. Ver. Bremen 18:143–174, Pl. 11.

Lemmermann, E. 1907. Brandenburgische Algen. IV. *Gonyaulax palustris* Lemm., eine neue Süßwasser-Peridinee. Beih. Bot. Centralblatt. 2, Abt., Syst., Pflanzengeogr., etc. 2:296–300.

Lemmermann, E. 1910. Kryptogamenflora der Mark Brandenburg, III, Algen I. Gebruder Borntraeger, Leipzig, 497–712.

Levander, K.M. 1892. Notizen über die Täfelung der Swchalenmembran des *Glenodinium cintum* Ehbg. Zool. Anz. 15:405–408.

Levander, K.M. 1894. Materialien zur Kenntnis der Wasserfauna in der Umgebung von Helsingfors mit besonder Berücksichigung der Meeresfauna. I. Protozoa. Acta Soc. Fauna Flora Fenn. 12 (2): 1–115. [*Glenodinium balticum*]

Levander, K.M. 1900a. Zur Kenntnis des Lebans in den stehenden Kleingewässern auf den Skäreninseln. Acta Soc. Fauna Flora Fenn. 18 (6): 1–107. [original for *Hemidinium ochraceus*]

Levander, K.M. 1900b. Zur Kenntnis der Fauna und Flora finnischer Binnenseen. Acta Soc. Fauna Flora Fenn. 19:4–55. [lists of organisms]

Levander, K.M. 1902. Eine neue farblose *Peridinium*-Art, *Peridinium achromaticum* n. sp. Medd. Soc. Fauna Flora Fenn. 28:49–50.

Levy, M.G., R.W. Litaker, R.J. Goldstein, M.J. Dykstra, M.W. Vandersea, and E.J. Noga. 2007. *Piscinoodinium*, a fish-ectoparasitic dinoflagellate, is a member of the class Dinophycease, subclass Gymnodiniphycidae: convergent evolution with *Amyloodinium*. J. Parasitol. 93 (5): 1006–1015.

Lindberg, K., Ø. Moestrup, and N. Daugbjerg. 2005. Studies on woloszynskoid dinoflagellates 1: *Woloszynskia coronata* re-examined using light and electron microscopy and partial LSU rDNA sequences, with description of *Tovellia* gen. nov. and *Jadwigia* gen. nov. (Tovelliaceae fam.nov.). Phycologia 44 (4): 416–440.

Lindemann, E. 1918. Untersuchungen über Süßwasserperineen und ihre Variationsformen II. Arch. Naturgesch. 1918 A 8:121–194.

Lindemann, E. 1925. Klasse: Dinoflagellatae. From Eyferth-Schoenichen's "Einfachste Lebensformen" Berlin 1:144–195.

Lindemann, E. 1928. Neue Peridineen. Hedwigia 68:291–296.

Lindemann, E. 1929. Experimentelle Studien über die Fortpflanzungser-scheinungen der Süsswasserperidineen auf Grund von Reinkulturen. Arch. Prostistenkd 68:1–104.

Lindley, J. 1846. The vegetable kingdom or, the structure, classification, and uses of plants, illustrated upon the natural system. Bradbury & Evans, London.

[Litvinenko] Lytvynenko, R.M. 1973. Analiz vydovogo skladu perydynyjevych vodorostej Ukrajiny. (Analysis of species composition of Peridineae in the Ukraine.) Ukraïns'kyi botanichnyi zhurnal. 30 (2): 169–174.

[Litvinenko] Lytvynenko, R.M. 1977. [possible transfer of *Gymnodinium bohemicum* to *Katodinium*, unable to confirm].

Loeblich, A.R., III. 1965. Dinoflagellate nomenclature. Taxon 14 (1): 15–18.

Loeblich, A.R., III. 1967. Notes on the divisions Chlorophyta, Chrysophyta, Pyrrhophyta, and Xanthophyta and the family Paramastigaceae. Taxon 16:230–236.

Loeblich, A.R., III. 1969. The amphiesma or dinoflagellate cell covering. Proceedings of the North American Paleontological Convention, Part G, 867–929.

Loeblich, A.R., III. 1980. Dinoflagellate nomenclature. Taxon 29:321–323.

Loeblich, A.R., Jr., and Loeblich, A.R., III. 1966. Index to genera, subgenera, and sections of the Pyrrhophyta. Studies in Tropical Oceanography, Miami 3.

Logares, R., K. Shalchian-Tabrizi, A. Boltovskoy, and K. Renefors. 2007a. Extensive dinoflagellate phylogenies indicate infrequent marine-freshwater transitions. Mol. Phylogen. Evol. 45:887–903.

Logares, R., K. Renefors, A. Kremp, K. Shalchian-Tabrizi, A. Boltovskoy, T. Tengs, A. Shurtleff, and D. Klaveness. 2007b. Phenotypically different microalgal morphospecies with identical ribosomal DNA: a case of rapid adaptive evolution? Microb. Ecol. 53:549–561.

Lom, J. 1981. Fish invading dinoflagellates: a synopsis of existing and newly proposed genera. Folia Parasitol. (Praha) 28:3–11.

Lom, J., and G. Schubert. 1983. Ultrastructural study of *Piscinoodinium pillulare* (Schäperclaus, 1954) Lom, 1981 with special emphasis on its attachment to the fish host. J. Fish Dis. 6:411–428.

Lopez-Lopez, E., and J.A. Serna-Hernandez. 1999. Variación estacional del zoo-plankton del embalse Ignacio Allende, Guanajuato, México y su relación con el fitoplancton y factores ambientales. Rev. Biol. Trop. 47:643–657.

Lowe, C.W. 1927. Some freshwater algae of southern Québec. Trans. R. Soc. Can. 21:291–316.

Lowe, C.W. 1931. The algal flora of Spider Lake, Vancouver Island. Biol. Board Can. Annu. Rep. 1931:76.

Margalef, R. 1947. Algas de agua dulce de la laguna de Ariguanabo (Isla de Cuba). Publ. Inst. Biol. Apl. (Barcelona) 4:79–89.

Marshall, H. 1976. The phytoplankton of Lake Drummond, Dismal Swamp, Virginia. Castanea 41:1–9.

Marshall, H.G., and L. Burchardt. 2004. Phytoplankton composition within the tidal freshwater-oligohaline regions of the Rappahannock and Pamunkey rivers in Virginia. Castanea 69:272–283.

Martin, M.R. accessed 2006. www.cedareden.com/micr/dino.html

Maskell, W.M. 1887. On the fresh-water Infusoria of the Wellington District. Trans. N.Z. Inst. 20:3–19.

McIntire, C.D., G.L. Larson, and R.E. Truitt. 2007. Seasonal and interannual variability in the taxonomic composition and production dynamics of phytoplankton assemblages in Crater Lake, Oregon. Hydrobiologia 574 (1): 179–204.

Meunier, S. 1919. Microplankton de la Mer Flamande III. Les Péridiniens. Mem. Mus. R. Hist. Nat. Belg. 8:1–116.

Meyer, B., and M. Elbrächter. 1996. (1235) Proposal to conserve the name *Peridinium elpatiewskyi* (Dinophyceae) with a conserved type. Taxon 45:531–532.

Meyer, R.L. 1969. The freshwater algae of Arkansas. Arkansas Acad. Sci. Proc. 23:145–156.

Meyer, R.L., and A.J. Brook. 1969. Freshwater algae from the Itasca State Park, Minnesota. III. Pyrrhophyta and Euglenophyta. Nova Hedwigia 18:369–382.

Meyer, R.L, J.H. Wheeler, and J.R. Brewer. 1970. The freshwater algae of Arkansas II. New Additions. Arkansas Acad. Sci. Proc. 24:32–35.

Moestrup, Ø., G. Hansen, and N. Daugbjerg. 2008. Studies on woloszynskoid dinoflagellates III: on the ultrastructure and phylogeny of *Borghiella dodgei* gen. et sp nov., a cold-water species from Lake Tovel, N. Italy, and on *B-tenuissima* comb. nov (syn. *Woloszynskia tenuissima*). Phycologia 47 (1): 54–78.

Moore, J.W. 1981. Seasonal abundance of *Ceratium hirundinella* (O.F. Müller) Schrank in lakes of different trophy. Arch. Hydrobiol. 92 (4): 535–548.

Morgan, G.R. 1971. Phytoplankton productivity vs dissolved nutrient level of Flathead Lake, Montana. PhD dissertation, Univ. Utah, Salt Lake City.

Morling, G. 1979. *Ceratium carolinianum* and the related *Ceratium carolinianum* var. *elongatum,* and *Ceratium cornutum.* A study of the distribution and morphology. Nova Hedwigia 31:937.

Morris, F.A., M.K. Morris, L.R. Williams, W.D. Taylor, F.A. Hiatt, S.C. Hern, J.W. Hilgert, and V.W. Lambou. 1978. Distribution of phytoplankton in Georgia lakes. EPA-600/3-78-011.

Morris, F.A., M.K. Morris, W.D. Taylor, L.R. Williams, S.C. Hern, and V.W. Lambou. 1979a. Distribution of phytoplankton in Washington lakes. EPA-600/3-79-121.

Morris, F.A., M.K. Morris, W.D. Taylor, L.R. Williams, S.C. Hern, and V.W. Lambou. 1979b. Distribution of phytoplankton in Nebraska lakes. EPA-600/3-79-066.

Morris, F.A., M.K. Morris, W.D. Taylor, L.R. Williams, S.C. Hern, and V.W. Lambou. 1979c. Distribution of phytoplankton in Idaho lakes. EPA-600/3-79-115.

Morris, M.K., L.R. Williams, W.D. Taylor, F.A. Hiatt, S.C. Hern, J.W. Hilgert, V.W. Lambou, and F.A. Morris. 1978a. Distribution of phytoplankton in Illinois lakes. EPA-600/3-79-050.

Morris, M.K., L.R. Williams, W.D. Taylor, F.A. Hiatt, S.C. Hern, J.W. Hilgert, V.W. Lambou, and F.A. Morris. 1978b. Distribution of phytoplankton in North Carolina lakes. EPA-600/3-78-051.

Morris, M.K., W.D. Taylor, L.R. Williams, S.C. Hern, V.W. Lambou, and F.A. Morris. 1979a. Distribution of phytoplankton in Missouri lakes. EPA-600/3-79-065.

Morris, M.K., W.D. Taylor, L.R. Williams, S.C. Hern, V.W. Lambou, and F.A. Morris. 1979b. Distribution of phytoplankton in California lakes. EPA-600/3-79-113.

Morris, M.K., W.D Taylor, L.R. Williams, S.C. Hern, V.W. Lambou, and F.A. Morris. 1979c. Distribution of phytoplankton in Colorado lakes. EPA-600/3-79-114.

Müller, O.F. 1773. Vermium terrestrium et fluviatilium seu animalium infusoriorum, helminthicorum et testaceorum, non marinorum secincta historia. Heineck et Faber typos Martini Hallager, Havniae et Lipsiae. Vol. 1.

Müller, O. F. 1786. Animalcula Infusoria fluviatilia et marina. Opus posthumum cura. O. Fabricii. Nicol. Möller, Havniae.

Murphy, T., B. Naples, D. Morgan, C. Dietz, A. Shurtleff, and J.R. Holt. 1994. Attachment of freshwater dinoflagellates relative to substrate orientation. J. Pa. Acad. Sci. 67:189

Murray, S., M.F. Jørgensen, N. Daugbjerg, and L. Rhodes. 2004. *Amphidinium* revisited. II. Resolving species boundaries in the *Amphidinium operculatum* species complex (Dinophyceae), including the descriptions of *Amphidinium trulla* sp. nov. and *Amphidinium gibbosum*. comb. nov. J. Phycol. 40: 366–382.

Netzel, H., and G. Dürr. 1984. Dinoflagellate cell cortex. In D.L. Spector, ed., Dinoflagellates. Academic, NY, 43–105.

Nicholls, K.H. 1973. Observations on red colored cells of *Peridinium wisconsinense* Eddy from Buckhorn Lake, Ontario. Trans. Am. Microsc. Soc. 92 (3): 526–528.

Nicholls, K.H. 1998. *Amphidiniopsis sibbaldii* sp. nov. (Thecadiniaceae, Dinophyceae), a new freshwater sand-dwelling dinoflagellate. Phycologia 37:334–339.

Nicholls, K.H. 1999. Validation of *Amphidiniopsis sibbaldii* Nicholls (Dinophyceae). Phycologia 38 (1): 74.

Nicholls, K.H., W. Kennedy, and C. Hammett. 1980. A fish-kill in Heart Lakes, Ontario, associated with the collapse of a massive population of *Ceratium hirundinella* (Dinophyceae). Freshw. Biol. 10:553–561.

Nolen, S.L., J.H. Carroll, D.L. Combs, J.C. Staves, and J.N. Veenstra. 1989. Limnology of Tenkiller Ferry Lake, Oklahoma, 1985–1986. Proc. Okla. Acad. Sci. 69:45.

Nygaard, G. 1926. Plankton from lakes of the Malayan region. Videnskabelige Meddelelser fra Dansk Naturhistorisk Forening i København 82:197–240.

Nygaard, G. 1945. Dansk Plante-Plankton. København. Gyldendalske Boghandel Nordisk Forlag. 1–52, Pl. 1–4.

Nygaard, G. 1949. Hydrobiological studies on some Danish ponds and lakes. Part II. The quotient hypothesis and some new or little known (plankton) organisms. Det Kongelige Danske Videnskabernes Selskab Biologiske Skrifter 7 (1): 1–293 (126 figures).

Olrik, K. 1992. Ecology of *Peridinium willei* and *P. volzii* (Dinophyceae) in Danish lakes. Nord. J. Bot. 12:557–568.

Ortega, M.M. 1984. Catálogo de algas continentales recientes de México. Univ. Nac. Autón. de México.

Oshima, Y., H. Minami, Y. Takano, and T. Yasumoto. 1989. Ichthyotoxins in a freshwater dinoflagellate *Peridinium polonicum*. In T. Okaichi, D.H. Anderson and T. Nemoto, eds., Red tides: biology environmental science and toxiciology. Proceedings of the First International Symposium on Red Tides. Elsevier, NY, 375–378.

Osorio-Tafall, B.F. 1942. Estudios sobre el plancton de Mexico 1. El genero *Lophodinium* Lemm. (Dinophyceae Peridiniales). Ciencia, Rev. Hispanoam. Cienc. Apl. 3:111–119. [also see listing under Tafall]

Ostenfeld, C.H. 1903. Phytoplankton from the sea around the Faeröes. In: Botany of the Faeröes based upon Danish investigations, Part II. Copenhagen, Det Nordiske Forlag, 558–612. [transfers *Prorocentrum compressum* to *Exuviaella*]

Ostenfeld, C.H. 1907. Beiträge zur Kenntnis der Algenflora des Kossogol-Beckens in der nordwestlichen Mongolei, mit spezieller Berücksichtigung des Phytoplanktons. Hedwigia 46:365–420.

Ostenfeld, C.H. 1908. The phytoplankton of the Aral Sea and its effluents. Wissenschaftliche Ergebnisse der Aralsee-Expedition. St Petersburg 8:123–225.

Ostenfeld, C.H., and G. Nygaard. 1925. On the phytoplankton of the Gatun Lake, Panama Canal. Dansk Bot. Arkiv 4 (10): 1–16.

Park, H.-D., and H. Hayashi. 1993. Role of encystment and excystment of *Peridinium bipes* f. *occulatum* (Dinophyceae) in freshwater red tides in Lake Kizaki, Japan. J. Phycol. 29:435–441.

Parson, M.J., and B.C. Parker. 1989. Algae flora in Mountain Lake, Virginia: past and present. Castanea 54 (2): 79–86.

Pascher, A. 1916a. Über eine neue Amöbe (*Dinamoeba varians*) mit dinoflagellatenartigen Schwärmern. Arch. Protistenkd. 36:118–136, Pl. 10.

Pascher, A. 1916b. Fusionplasmodien bei Flagellaten und ihre Bedeutung für die Ableitung der Rhizopoden von den Flagellaten. Arch. Protistenkd. 37 (1): 31–64.

Pascher, A. 1927. Die braune Algenreihe aus der Verwandtschaft der Dinoflagellaten (Dinophyceen). Arch. Protistenkd. 58:1–54.

Pascher, A. 1928. Von einer neuen Dinococcale (*Cystodinium phaseolus*) mit zwei verschiedenen Schwärmertypen. Arch. Protistenkd. 63:241–254.

Pascher, A. 1944a. Über neue, protococcoide, festsitzende Algengattungen aus der Verwandtschaft der Dinoflagellaten. Beih. Bot. Centralblatt 57 A (3): 376–395.

Pascher, A. 1944b. Über zwei farblose, protococcoide Algen aus der Reihe der Dinophyceen und über Schädigungen durch Epiphytimus. Beih. Bot. Centralblatt 57 A (3): 396–404.

Penard, E. 1891. Les Péridiniacées du Léman. Bull. Trav. Soc. Bot. Genève 6:1–63, Pl. 1–5.

Perez, E.A.A., J. DeCosta, and K.E. Havens. 1994. The effects of nutrient addition and pH manipulation in bag experiments on the phytoplankton of a small acidic lake in West Virginia, USA. Hydrobiologia 291:93–103.

Pfiester, L.A. 1971. Periodicity of *Ceratium hirundinella* (O.F.M.) Dujardin and *Peridinium cinctum* (O.F.M.) Ehrenberg in relation to certain ecological factors. Castanea 36:246–257.

Pfiester, L.A. 1975. Sexual reproduction of *Peridinium cinctum* f. *ovoplanum* (Dinophyceae). J. Phycol. 11:259–265.

Pfiester, L.A. 1976. Sexual reproduction of *Peridinium willei* (Dinophyceae). J. Phycol. 12:234–238.

Pfiester, L.A. 1977. Sexual reproduction of *Peridinium gatunense* (Dinophyceae). J. Phycol. 13:92–95.

Pfiester, L.A. 1984. Sexual reproduction. In D. L. Spector, ed., Dinoflagellates. Academic, NY, 181–199.

Pfiester, L.A., and D.M. Anderson. 1987. Dinoflagellate reproduction. In F.J.R. Taylor, ed., The Biology of Dinoflagellates. Bot. Monogr. 21. Blackwell, Boston

Pfiester, L.A., and J.F. Highfill. 1993. Sexual reproduction of *Hemidinium nasutum* alias *Gloeodinium montanum*. Trans. Am. Microsc. Soc. 112:69–74.

Pfiester, L.A., and R.A. Lynch. 1980. Amoeboid stages and sexual reproduction of *Cystodinium bataviense* and its similarity to *Dinococcus* (Dinophyceae). Phycologia 19:178–183.

Pfiester, L.A., and J. Popovsky. 1979. Parasitic, amoeboid dinoflagellates. Nature 279:421–424.

Pfiester, L.A., and J.J. Skvarla. 1979. Heterothallism and thecal development in the sexual life history of *Peridinium volzii* (Dinophyceae). Phycologia 18:13–18

Pfiester, L.A., and J.J. Skvarla. 1980. Comparative ultrastructure of vegetative and sexual thecae of *Peridinium limbatum* and *Peridinium cinctum* (Dinophyceae). Am. J. Bot. 67:955–958.

Pfiester, L.A., and S. Terry. 1978. Additions to the algae of Oklahoma. Southwest. Nat. 23 (1): 85–94.

Pfiester, L.A., R.A. Lynch, and J.J. Skvarla. 1980. Occurrence, growth, and SEM portrait of *Woloszynskia reticulata* Thompson (Dinophyceae). Trans. Am. Microsc. Soc. 99:213–217.

Pfiester, L.A., P. Timpano, J.J. Skvarla, and J.R. Holt. 1984. Sexual reproduction and meiosis in *Peridinium inconspicuum* Lemmermann *(Dinophyceae)*. Am. J. Bot. 71:1121–1127.

Phillips, K.A., and M.W. Fawley. 2002. Winter phytoplankton community structure in three shallow temperate lakes during ice cover. Hydrobiologia 470:97–113.

Pienaar, R.N. 1980. The ultrastructure of *Peridinium balticum* (Dinophyceae) with particular reference to its endosymbionts. Electron Microsc. Soc. South. Afr. Proc. 10:75–76.

Playfair, G.I. 1912. Plankton of the Sydney water supply. Proc. Linn. Soc. N.S.W. 37:512–552.

Playfair, G.I. 1919. Peridineae of New South Wales. Proc. Linn. Soc. N.S.W. 44:793–818.

Pollinger, U. 1987. Freshwater ecosystems. In F.J.R. Taylor, ed., The Biology of Dinoflagellates. Blackwell, Malden, MA, 502–529.

Popovský, J. 1970. Some thecate dinoflagellates from Cuba. Arch. Protistenkd. 112:252–258.

Popovský, J. 1971a. Some remarks to the life cycle of *Gloeodinium montanum* KLEBS and *Hemidinium nasutum* STEIN (Dinophyceae). Arch. Protistenkd. 113: S.131–136.

Popovský, J. 1971b. Some interesting freshwater dinoflagellates from central Europe. Arch. Protistenkd. 113: S.277–284.

Popovský, J. 1982. Another case of phagotrophy by *Gymnodinium helveticum* Penard f. *achroum* Skuja. Arch. Protistenkd. 125:73–78.

Popovský, J. 1985. *Gymnodinium austriacum* Schiller (Dinophyceae), a case of phagotrophy. Algol. Stud. 40, Arch. Hydrobiol. Suppl. 71 (3): 425–429.

Popovský, J., and L.A. Pfiester. 1982. The life-histories of *Stylodinium sphaera* Pasher and *Cystodinedria inermis* (Geitler) Pascher (Dinophyceae), two freshwater facultative predator-autotrophs. Arch. Protistenkd. 125:115–127.

Popovský, J., and L.A. Pfiester. 1986. A taxonomical note to the section Umbonatum of the genus *Peridinium* Ehrenberg, 1932 (Dinophyceae). Arch. Protistenkd. 132:73–77.

Popovský, J., and L.A. Pfiester. 1990. Süßwasserflora von Mitteleuropa. Vol. 6: Dinophyceae (Dinoflagellida). Gustav Fischer Verlag, Jena.

Pouchet, G. 1887. Quatrieme contribution à l'histoire des Peridiniens. Journal de l'anatomie et de la Physiologie normales et pathologiques de l'homme et des animaux. Paris 23:87–112, Pl. 9, 10.

Poulin, M., P.B. Hamilton, and M. Proulx. 1995. Catalogue des algues d'eau douce du Québec, Canada. Can. Field-Nat. 109:27–110.

Prager, J.C. 1963. Fusion of the family Glenodiniaceae into the Peridiniaceae, with notes on *Glenodinium foliaceum* Stein. J. Protozool. 10 (2): 195–204.

Prescott, G.W. 1927. The motile algae of Iowa. In Aquatic flora in Iowa. Univ. Iowa Stud. Nat. Hist. 12 (6).

Prescott, G.W. 1944. New species and varieties of Wisconsin algae. Farlowia 1 (3): 347–385.

Prescott, G.W. 1951a. Ecology of Panama Canal algae. Trans. Am. Microsc. Soc. 70:1–24.

Prescott G.W. 1951b. Algae of the western Great Lakes area exclusive of desmids and diatoms. Cranbrook Inst. Sci. Bull. 31.

Prescott, G.W. 1955. Algae of the Panama Canal and its tributaries I. Flagellated organisms. Ohio J. Sci. 55 (2): 99–121.

Prescott, G.W., and H.T. Croasdale. 1937. New or noteworthy fresh water algae of Massachusetts. Trans. Am. Microsc. Soc. 56 (3): 269–282.

Prescott, G.W., and G.E. Dillard. 1979. A checklist of algal species reported from Montana 1891–1977. Monogr. 1, Montana Acad. Sci. 38.

Prescott, G.W., and W.C. Vinyard. 1965. Ecology of Alaskan freshwater algae V. Limnology and flora of Malikpuk Lake. Trans. Am. Microsc. Soc. 84:427–478.

Prescott, G.W., H. Silva, and W.E. Wade. 1949. New or otherwise interesting fresh-water algae from North America. Hydrobiologia 2 (1): 84–93.

Puytorac, P. de, J.P. Mignot, J. Grain, C.A. Grolière, L. Bonnet, and P. Couillard. 1972. Premier relevé de certains groupes de protozoaires libres sur le territoire de las Station de biologie de l'Université de Montréal (Saint-Hippolyte, comté de Terrebonne, Québec). Nat. Can. 99:417–440.

Raymond, M.R. 1937. A limnological study of the plankton of a concretion-forming marl lake. Trans. Am. Microsc. Soc. 56 (4): 405–430.

Reimchen, T.E., and J. Buckland-Nicks. 1990. A novel association between an endemic stickleback and a parasitic dinoflagellate. 1. Seasonal cycle and host response. Can. J. Zool. 68:667–671.

Reisser, W. 1992. Endosymbiotic associations of algae with freshwater protozoa and invertebrates. In W. Reisser, ed., Algae and symbioses. Biopress, 1–20.

Reisser, W., and M. Widowski. 1992. Taxonomy of eucaryotic algae endosymbiotic in freshwater associations. In W. Reisser, ed., Algae and symbioses. Biopress, 21–40.

Rengefors, K. 1998. Seasonal succession of dinoflagellates coupled to the benthic cyst dynamics in Lake Erken. Arch. Hydrobiol., Spec. Issues, Adv. Limnol. 51:123–141.

Rengefors, K., and C. Legrand. 2001. Toxicity in *Peridinium aciculiferum*: an adaptive strategy to outcompete other winter phytoplankton? Limnol. Oceanogr. 46 (8): 1990–1997.

Richards, B.C. 1962. The morphology, cytology and life history of *Urococcus insignis* (Hass.) Kutz. MS thesis, Univ. Kans.

Roset, J., A. Gibello, S. Aguayo, L. Dominguez, M. Álvarez, J.F. Fernández-Garayzabal, A. Zapata, and M.J. Muñoz. 2002. Mortality of rainbow trout [*Oncorynchus mykiss* (Walbaum)] associated with freshwater dinoflagellate bloom [*Peridinium polonicum* (Woloszynska)] in a fish farm. Aquac. Res. 33:141–145.

Rushforth, S.R., and L.E. Squires. 1985. New records and comprehensive list of the algal taxa of Utah Lake, Utah, USA. Great Basin Nat. 45 (2): 237–254.

Saldarriaga, J.F., F.J.R. Taylor, T. Calalier-Smith, S. Menden-Deuer, and P.J. Keeling. 2004. Molecular data and the evolutionary history of dinoflagellates. Eur. J. Protistol. 40:85–111.

Saunders, G.W., D.R.A. Hill, J.P. Sexton, and R.A. Andersen. 1997. Small-subunit ribosomal RNA sequences from selected dinoflagellates: testing classical evolutionary hypotheses with molecular systematic methods. In D. Bhattacharya, ed., Origins of Algae and their plastids. Springer Wien, NY.

Schäperclaus, W. 1951. Der Colisa-Parasit, einneuer Krankheitserreger bei Aquarienfischen. Aquar.-Terr.-Z. 4:169–171.

Schäperclaus, W. 1954. Handbuch der Fischkrankheite. Akademie Verlag, Berlin.

Schiller, J. 1933. Dinoflagellatae (Peridineae). In R. Kolkwitz, ed., Rabenhorst's KryptogamenFlora von Deutschland, Österreich und der Schweiz, 1. Teil. Akademische Verlag, Leipzig.

Schiller, J. 1935/1937. Dinoflagellatae (Peridineae). In R. Kolkwitz, ed., Rabenhorst's KryptogamenFlora von Deutschland, Österreich und der Schweiz, 2. Teil. Akademische Verlag, Leipzig.

Schiller, J. 1954. Über winterliche pflanzliche Bewohner des Wassers, Eises und des darauf liegender Schneebreies. I. Österr. Bot. Z. 101:236–284. [*Gymnodinium pasheri*]

Schiller, J. 1955. Untersuchungen an den planktischen Protophyten des Neusiedlersees 1950–1954. Wiss. Arb. Burgenland 9:1–66.

Schilling, A.J. 1891a. Die Süßwasser-Perideen. Flora Allg. Bot. Z. 74:220–299, Pl. 8–10.

Schilling, A.J. 1891b. Kleiner Breitrag zur Technik der Flagellatenforschung. Z. Wiss. Mikrosk. 8:314.

Schilling, A.J. 1891c. Untersuchungen über die thierische Lebensweise einiger Peridineen. Ber. Deutsch. Bot. Ges. 9:199–208+, Pl. 1–17. [*Glenodinium edax*]

Schilling, A.J. 1913. Dinoflagellatae (Peridineae). In A. Pascher, ed., Süßwasser-flora, vol. 3. Gustav Fischer, Jena, 1–66.

Schindler, D.W., and S.K. Holmgren. 1971. Primary production and phytoplankton in the Experimental Lakes Area, northwestern Ontario, and other low-carbonate waters, and a liquid scintillation method for determining ^{14}C activity in photosynthesis. J. Fish. Res. Board Can. 28:189–201.

Schmitter-Soto, J.J., F.A. Comín, E. Escobar-Briones, J. Herrera-Silveira, J. Alcocer, E. Suárez-Morales, M. Elías-Guriérrez, V. Díaz-Arce, L.E. Marín, and B. Steinich. 2002. Hydrogeochemical and biological characteristics of cenotes in the Yucatan Peninsula (SE Mexico). Hydrobiologia 467:215–228.

Schrank, F.P. 1793. Mikroskopische Wahrnehmungen. Der Naturforscher 27:26–37.

Schröder, B. 1918. Die neun wesentlichen Formentypen von *Ceratium hirundinella* O.F. Muller. Arch. Naturgeschichte 84:222–230.

Schumacher, G. 1956. A qualitative and quantitative study of the plankton in southwestern Georgia ponds. Am. Midl. Nat. 56:88–115.

Schumacher, G. 1961. Biology of the Allegany Indian Reservation and vicinity. Part 1: The algae. Bull. 383, New York State Museum and Science Service, Albany, 5–18.

Schumacher, G.J., and W.C. Muenscher. 1952. Plankton algae of some lakes of Whatcom County, Washington. Madroño 2:289–297.

Seaborn, D.W., J.R. Merrell, and J.R. Holt. 1992. Effects of increased diversity and change in acid neutralizing capacity (ANC) on three *Peridinium* species of an acid sensitive lake in Union County, Pennsylvania. J. Pa. Acad. Sci. 65:199.

Shamess, J.J., G.G.C. Robinson, and L.G. Goldsborough. 1985. The structure and comparison of periphytic and plankton algal communities in two eutrophic prairie lakes. Arch. Hydrobiol. 103:99–116.

Sheath, R.G., and J.A. Hellebust. 1978. Comparison of algae in the euplankton, tycoplankton, and periphyton of a tundra pond. Can. J. Bot. 56:1472–1483.

Sheath, R.G., and A.D. Steinman. 1982. A checklist of freshwater algae of the Northwest Territories, Canada. Can. J. Bot. 60:1964–1997.

Sheath, R.G., M. Havas, J.A. Hellebust, and T.C. Hutchinson. 1982. Effects of long-term natural acidification on the algal communities of tundra ponds at the Smoking Hills, N.W.T., Canada. Can. J. Bot. 60:58–72.

Sicko-Goad, L., and G. Walker. 1979. Viroplasm and large virus-like particles in the dinoflagellate *Gymnodinium uberrimum*. Protoplasma 99:203–210.

Sieminska, J. 1965. Algae from Mission Wells Pond, Montana. Trans. Am. Microsc. Soc. 84:98–126.

Skuja, H. 1937. Botanische Ergebnisse der Expedition der Akademie der Wissenschaften in Wien nach Südwest-China 1914–1918. I. Teil: Algae. Symbolae Sinicae, Wien.

Skuja, H. 1939. Beitrag zur Algenflora von Lettland II. Acta Horti Bot. Univ. Latv. 11/12:41–169.

Skvortzov, B.W. 1927. Über *Gymnodinium hiemale* Skvortz. Hedwigia 67:122–124. [*Gymno. skvortzowii*]

Smith, G.M. 1933. The fresh-water algae of the United States. McGraw Hill.

Smith, G.M. 1950. The fresh-water algae of the United States, 2nd ed. McGraw Hill.

Smith, M.W. 1961. A limnological reconnaisance of a Nova Scotian brownwater lake. J. Fish. Res. Board Can. 18:463–478.

Spector, D.L., ed. 1984. Dinoflagellates. Academic, NY.

Spector, D.L., L.A. Pfiester, and R.E. Triemer. 1981. Ultrastructure of the dinoflagellate *Peridinium cinctum* f. *ovoplanum*. II. Light and electron microscopic observations on fertilization. Am. J. Bot. 68 (1): 34–43.

Spero, H.J. 1979. A study of the life-cycle, feeding and chemosensory behavior in the holozoic dinoflagellate *Gymnodinium fungiforme* Anissimova. MS thesis, Texas A&M Univ.

Starmach, K. 1974. Flora Slodkowodna Polski 4: Cryptophyceae, Dinophyceae, Raphidiophyceae. Panstwowe Wydawnictwo Naukowe. Warszwa, Kraków.

Steidinger, K.A., J.M. Burkholder, H.B. Glasgow, C.W. Hobbs, J.K. Garrett, E.W. Truby, E.J. Noga, and S.A. Smith. 1996a. *Pfiesteria piscicida* gen. et sp. nov. (Pfiesteriaceae fam. nov.), a new toxic dinoflagellate with a complex life cycle and behavior. J. Phycol. 32:157–164.

Steidinger, K.A., J.H. Landsberg, E.W. Truby, and B.A. Blakesley. 1996b. The use of scanning electron microscopy in identifying small "gymnodinioid" dinoflagellates. Nova Hedwigia 112:415–422.

Stein, F.R. von. 1878. Der organismus der Infusionsthiere nach eigenen Forschungen in systematischer Reihenfolge bearbeitet.III. Abteilung. Die Naturegeschichte der Flagellaten oder Geisselinfusorien. I. Hälfte. W. Engelmann, Leipzig.

Stein, F.R. 1883. Der organismus der Infusionsthiere nach eigenen Forschungen in systematischer Reihenfolge bearbeitet. III. Abteilung. II Hälfte. Die Naturegeschichte der arthrodelen Flagellaten. Verlag von Wilhelm Engelmann, Leipzig.

Stein, J.R. 1975. Freshwater algae of British Columbia: the lower Farser Valley. Syesis 8:119–184.

Stein, J.R., and C.A. Borden. 1979. Checklist of freshwater algae of British Columbia. Syesis 12:3–39.

Stewart, A.J., and D.W. Blinn. 1976. Studies on Lake Powell, USA: environmental factors influencing phytoplankton success in a high desert warm monomictic lake. Arch. Hydrobiol. 78 (2): 139–164.

Stoecker, D.K. 1998. Conceptual models of mixotrophy in planktonic protists and some ecological and evolutionary implications. Eur. J. Protistol. 34:281–290.

Stokes, A.C. 1887. Notices of new fresh-water infusoria. Proc. Am. Philos. Soc. 24 (126): 244–255.

Stokes, A.C. 1888. A preliminary contribution toward a history of the fresh-water infusioria of the United States. J. Trenton Nat. Hist. Soc. 1 (3): 141–145, Pl. 4, Fig. 1.

Stokes, P.M., and Y.K. Yung. 1986. Phytoplankton in selected La Cloche (Ontario) lakes, pH 4.2–7.0, with special reference to algae as indicators of chemical characteristics. In J.P. Smol, R. Batterbee, R.B. Davis, and J. Merilainen, eds., Diatoms and lake acidity. Dr. W. Junk, Dordrecht.

Stoneburner, D.L., and L.A. Smock. 1980. Plankton communities of an acidic, polymictic, brown-water lake. Hydrobiologia 69 (1–2): 131–137.

Stosch, H.A. von. 1973. Observations on vegetative reproduction and sexual life cycles of two freshwater dinoflagellates, *Gymnodinium pseudopalustre* Schiller and *Woloszynskia apiculata* sp. nov. Br. Phycol. J. 8:105–134.

Suchlandt, O. 1916. Dinoflagellaten als Erreger von rotem Schnee. Ber. Dtsch. Bot. Gesell. 34:242–246.

Suxena, M.R. 1983. Algae from Kodaikanal Hill, South India. Bibl. Phycol. 66: 43–99.

Tafall, B.F. Osorio. 1941. Materiales para el estudio del microplancton del Lago de Pátzcuaro (México). An. Escuela Nac. Cienc. Biol. 2:331–383.

Taft, C.E., and C.W. Taft. 1971. The algae of western Lake Erie. Bull. Ohio Biol. Survey 4 (1).

Takano, Y., G. Hansen, D. Fujita, and T. Horiguchi. 2008. Serial replacement of diatom endosymbionts in two freshwater dinoflagellates, *Peridiniopsis* spp. (Peridiniales, Dinophyceae). Phycologia 47 (1): 41–53.

Taylor, F.J.R., ed. 1987. The biology of dinoflagellates. Bot. Monogr. 21. Blackwell, Boston.

Taylor, W.D., F.A. Hiatt, S.C. Hern, J.W. Hilgert, V.W. Lambou, F.A. Morris, M.K. Morris, and L.R. Williams. 1977. Distribution of phytoplankton in Alabama lakes. EPA-600/3-77-082.

Taylor, W.D., F.A. Hiatt, S.C. Hern, J.W. Hilgert, V.W. Lambou, F.A. Morris, M.K. Morris, and L.R. Williams. 1978. Distribution of phytoplankton in Florida lakes. EPA-600/3-78-085.

Taylor, W.D., L.R. Williams, S.C. Hern, V.W. Lambou, F.A. Morris, and M.K. Morris. 1979a. Distribution of phytoplankton in Arizona lakes. EPA-600/3-79-112.

Taylor, W.D., L.R. Williams, S.C. Hern, V.W. Lambou, F.A. Morris, and M.K. Morris. 1979b. Distribution of phytoplankton in North Dakota lakes. EPA-600/3-79-067.

Taylor, W.D., L.R. Williams, S.C. Hern, V.W. Lambou, F.A. Morris, and M.K. Morris. 1979c. Distribution of phytoplankton in Oregon lakes. EPA-600/3-79-119.

Temponeras, M., J. Kristiansen, and M. Moustaka-Gouni. 2000. A new *Ceratium* species (Dinophyceae) from Lake Doïrani, Macedonia, Greece. Hydrobiologia 424:101–108.

Thomasson, K. 1962. Planktological notes from western North America. Ark. Bot. 4 (14): 437–463.

Thompson, R.H. 1938. A preliminary survey of the fresh-water algae of eastern Kansas. Univ. Kans. Sci. Bull. 25:5–83.

Thompson, R.H. 1947. Fresh-water dinoflagellates of Maryland. State of Maryland Board of Natural Resources 67: 3–24. Chesapeake Biol. Lab., Solomons, MD.

Thompson, R.H. 1949. Immobile Dinophyceae. I. New records and a new species. Am. J. Bot. 36:301–308.

Thompson, R.H. 1950. A new genus and new records of fresh water Pyrrophyta in the Desmokontae and Dinophyceae. Lloydia 13:277–299.

Thompson, R.H. 1959. Algae. In W.T. Edmondson, ed., H.B. Ward and G.C. Whipple, Fresh-water Biology, 2nd ed. John Wiley & Sons, NY, 115–170.

Thompson, R.H., and R.L. Meyer. 1984. *Cystodinium acerosum* n. sp. and certain "horned" species of the genus *Cystodinium* Klebs. Trans. Kans. Acad. Sci. 87 (3–4): 83–90.

Tiffany, L.H., and M.E. Britton. 1952. The algae of Illinois. Univ. Chicago Press.

Timpano, P., and L.A. Pfiester. 1985. Colonization of the epineuston by *Cystodinium bataviense* (Dinophyceae): behavior of the zoospore. J. Phycol. 21:56–62.

Timpano, P., and L.A. Pfiester. 1986. Cell layers of hypnozygotes in *Peridinium* (Dinophyceae). Trans. Am. Microsc. Soc. 105 (4): 381–386.

Tippit, D.H., and J.D. Pickett-Heaps. 1976. Apparent amitosis in the binucleate dinoflagellate *Peridinium balticum*. J. Cell Sci. 21:273–289.

Toetz, D., and J. Windell. 1993. Phytoplankton in a high-elevation lake, Colorado front range: application to lake acidification. Great Basin Nat. 53 (4): 350–357.

Tomas, R.N., and E.R. Cox. 1973. Observations on the symbiosis of *Peridinium balticum* and its intracellular alga. I. Ultrastructure. J. Phycol. 9:304–323.

Tomas, R.N., E.R. Cox, and K.A. Steidinger. 1973. *Peridinium balticum* (Levander) Lemmermann, an unusual dinoflagellate with a mesocaryotic and a eucaryotic nucleus. J. Phycol. 9 (1): 91–98.

Truby, E.W. 1997. Preparation of single-celled marine dinoflagellates for electron microscopy. Microsc. Res. Tech. 36:337–340.

Umaña Villalobos, G. 1988. Fitoplancton de las lagunas Barba, Fraijanes y San Joaquín, Costa Rica. Rev. Biol. Trop. 36 (2B): 471–477.

Van Meter-Kasanof, N. 1973. Ecology of the micro-algae of the Florida Everglades. Part I. Environment and some aspects of freshwater periphyton, 1959–1963. Nova Hedwigia 24:619–664.

Vinyard, W.C. 1957. Algae of the Glacier National Park region, Montana. Proc. Mont. Acad. Sci. 17:49–53.

Vinyard, W.C. 1958 The algae of Oklahoma (exclusive of diatoms). PhD thesis, Mich. State Univ.

Vyverman, W., and P. Compère. 1991. Electron microscopical study of some dinoflagellates (Dinophyta) from Papua New Guinea. Bull. Jard. Bot. Nat. Belg. 61: 269–277.

Wailes, G.H. 1928a. Dinoflagellates from British Columbia with descriptions of new species. Study from the stations of the Biological Board of Canada. Part I. With Plates 1–3. Mus. Art Notes, Vancouver 3 (1): 20–31. [*P. striolatum* for *P. vancouverense*]

Wailes, G.H. 1928b. Dinoflagellates from British Columbia with descriptions of new species. Part II. With Plates 4–6. Mus. Art Notes, Vancouver 3 (2): 27–35.

Wailes, G.H. 1934. Freshwater dinoflagellates of North America. Mus. Art Notes, Vancouver 7 Suppl. 11: 1–10, Pl. 1–4.

Wall, D., and W.R. Evitt. 1975. A comparison of the modern genus *Ceratium* Schrank, 1793, with certain Cretaceous marine dinoflagellates. Micropaleontology 21:14–44, Pl. 1–3.

Webber, E.E. 1961. A list of algae from selected areas in Massachusetts. Rhodora 63:275–281.

Wedemayer, G.J., and L.W. Wilcox. 1984. The ultrastructure of the freshwater, colorless dinoflagellate *Peridiniopsis berolinense* (Lemm.) Bourrelly (Mastigophora, Dinoflagellida). J. Protozool. 31:444–453.

Wedemayer, G.J., L.W. Wilcox, and L.E. Graham. 1982. *Amphidinium cryophilum* sp. nov. (Dinophyceae) a new freshwater dinoflagellate. I. Species description using light and scanning electron microscopy. J. Phycol. 18:13–17.

Wehr, J.D., and R.G. Sheath, eds. 2003. Freshwater algae of North America: ecology and classification. Academic.

West, G.S. 1907. Report on the freshwater algae, including phytoplankton, of the Third Tanganyika Expedition conducted by Dr. W. A. Cunnington, 1904–1905. Linn. J. Bot. 38:81–195. [10 plates]

West, W., and G.S. West. 1905. A further contribution to the freshwater plankton of the Scottish locks. Trans. R. Soc. Edinb. 41 (part 3, no. 2): 477–519.

Whelden, R.M. 1947. Algae. In N. Polunin, ed., Botany of the Canadian Eastern Arctic Part II: Thallophyta and Bryophyta. Natl. Mus. Can. Bull. 97.

Whitford, L.A., and Y.C. Kim. 1971. Algae from alpine areas in Rocky Mountain National Park, Colorado. Am. Midl. Nat. 85:425–430.

Whitford, L.A., and G.J. Schumacher. 1969. A manual of fresh-water algae in North Carolina. North Carolina Agricultural Experimental Station.

Whitford, L.A., and G.J. Schumacher. 1984. A manual of fresh-water algae. Sparks Press, Raleigh, NC.

Whiting, M.C., J.D. Brotherson, and S.R. Rushforth. 1978. Environmental interactions in summer algal communities of Utah Lake. Great Basin Nat. 38 (1): 31–41.

Wilcox, L.W., and G.J. Wedemayer. 1984. *Gymnodinium acidotum* Nygaard (Pyrrhophyta), a dinoflagellate with an endosymbiotic cryptomonad. J. Phycol. 20:236–242.

Wilcox, L.W., and G.J. Wedemayer. 1985. Dinoflagellate with blue-green chloroplasts derived from an endosymbiotic eukaryote. Science 227:192–194.

Williams, L.R., W.D. Taylor, F.A. Hiatt, S.C. Hern, J.W. Hilgert, V.W. Lambou, F.A. Morris, R.W. Thomas, and M.K. Morris. 1977. Distribution of phytoplankton in Mississippi lakes. EPA-600/3-77-101.

Williams, L.R., F.A. Morris, J.W. Hilgert, V.W. Lambou, F.A. Hiatt, W.D. Taylor, M.K. Morris, and S.C. Hern. 1978. Distribution of phytoplankton in New Jersey lakes. EPA-600/3-78-014.

Williams, L.R., S.C. Hern, V.W. Lambou, F.A. Morris, M.K. Morris, and W.D. Taylor. 1979a. Distribution of phytoplankton in Wyoming lakes. EPA-600/3-79-122.

Williams, L.R., S.C. Hern, V.W. Lambou, F.A. Morris, M.K. Morris, and W.D. Taylor. 1979b. Distribution of phytoplankton in Kansas lakes. EPA-600/3-79-063.

Williams, L.R., S.C. Hern, V.W. Lambou, F.A. Morris, M.K. Morris, and W.D. Taylor. 1979c. Distribution of phytoplankton in Iowa lakes. EPA-600/3-79-062.

Williams, L.R., S.C. Hern, V.W. Lambou, F.A. Morris, M.K. Morris, and W.D. Taylor. 1979d. Distribution of phytoplankton in Utah lakes. EPA-600/3-79-120.

Wołoszyńska, J. 1912. Das Phytoplankton einiger Javanischer Seen, mit Berücksichtigung des Sawa-Planktons. Bull. Int. Acad. Sci. Cracovie (B) 1912:649–709.

Wołoszyńska, J. 1913. Über die Süsswasseraten der Gatlung *Ceratium* Schrank. Kosmos 38:414–432.

Wołoszyńska, J. 1916. Polskie Peridineae słodkowodne.—Polnische Süßwasser-Peridineen. Bull. Int. Acad. Sci. Cracovie (B): 260–284, Pl. 10–14.

Wołoszyńska, J. 1917. Nowe gatunki Peridineów, tudzież spostrzeżenia nad budow? okrywy u Gymnodiniów I Glenodiniów.—Neue Peridineen-Arten, nebst Bemerkungen über den Bau der Hülle bei Gymno- und Glenodinium. Bull. Int. Acad. Sci. Cracovie (B): 114–122.

Wołoszyńska, J. 1919. Die Algen der Tatraseen und Tümpel I. Bull. Int. Acad. Sci Cracovie (B): 196–200 Plate 14.

Wołoszyńska, J. 1925a. Notatki algologiczne (Algologische Notizen). 1. *Amphidinium vigrense* n. sp. 2. *Peridinium* sp. Sprawozdania Stacji Hydrobiologicznej na Wigrach 1 (4): 1–9 [Also published in French under the title "Comptes Rendus de la Station hydrobiologique du lac de Wigry"]

Wołoszyńska, J. 1925b. Przyczynki do znajomości polskich bróznic słodkowodnych. (or: Beiträge zur Kenntnis der Süsswasser-Dinoflagellaten Polens.) Acta Soc. Bot. Poloniae 3 (1): 49–64. [*Gymnodinium undulatum*]

Wołoszyńska, J. 1928. Dinoflagellatae Polskiego Bałtyku i błot nad piaśnicą. Archiwum Hydrobiologji i Rybactwa 3 (3–4): 153–250, Pl. 2–15 (Polish version).

Wołowski, K., J. Piatek, and B.J. Płachno. 2011. Algae and stomatocytes associated with carnivorous plants. First report of chrysophyte stomatocysts from Virginia, USA. Phycologia 50 (5): 511–519.

Woodson, B.R., Jr. 1969. Algae of a fresh-water Virginia pond. Castanea 34: 352–374.

Woodson, B.R., and K. Seaburg. 1983. Seasonal changes in the phytoplankton of Lake Chesdin, Virginia with ecological observations. Va. J. Sci. 34:257–272.

Wright, R.T. 1964. Dynamics of a phytoplankton community in an ice-covered lake. Limnol. Oceanogr. 9:163–178.

Wu, J.-T., L.-L. Kuo-Huang, and J. Lee. 1998. Algicidal effect of *Peridinium bipes* on *Microcystis aeruginosa*. Curr. Microbiol. 37:257–261.

Wunderlin, T.F. 1971. A survey of the freshwater algae of Union County, Illinois. Castanea 36:1–53.

Yan, N.D., and P. Stokes. 1978. Phytoplankton of an acidic lake, and its responses to experimental alterations of pH. Environ. Conserv. 5 (2): 93–100.

Yeo, H.W. 1971 The composition, abundance, and seasonal periodicity of phytoplankton at Lake Winnipesaukee, New Hampshire. PhD dissertation, Univ. N.H., Durham.

Zederbauer, E. 1904a. *Ceratium hirundinella* in den österreichischen Alpenseen. Österr. Bot. Z. Jahrg. 54 (4): 124–128, Taf 5, Figs. 1–7.

Zederbauer, E. 1904b. *Ceratium hirundinella* in den österreichischen Alpenseen. Österr. Bot. Z. Jahrg. 54 (5): 167–172, Taf 5, Figs. 8–25.

Zohary, T., U. Pollinger, O. Hadas, and K.D. Hambright. 1998. Bloom dynamics and sedimentation of *Peridinium gatunense* in Lake Kinneret. Limnol. Oceanogr. 43 (2): 175–186.

Taxonomic Index

Numbers in **boldface** indicate main entries.
Numbers in *italics* indicate figures.
P indicates black-and-white plates.
CP indicates color plates.
An asterisk (*) indicates other names in the literature.